A MORAL TECHNOLOGY

A Moral Technology

Electrification as Political Ritual in New Delhi

Leo Coleman

Cornell University Press
Ithaca and London

First published 2017 by Cornell University Press
First printing, Cornell Paperbacks, 2017

Printed in the United States of America

Library of Congress Cataloging-in-Publication Data

Names: Coleman, Leo, author.
Title: A moral technology : electrification as political ritual in New Delhi / Leo Coleman.
Description: Ithaca : Cornell University Press, 2017. | Includes bibliographical references and index.
Identifiers: LCCN 2016043403 (print) | LCCN 2016046746 (ebook) | ISBN 9781501707513 (cloth : alk. paper) | ISBN 9781501707520 (pbk. : alk. paper) | ISBN 9781501707919 (epub/mobi) | ISBN 9781501707926 (pdf)
Subjects: LCSH: Electrification—Political aspects—India—New Delhi. | Electric utilities—Political aspects—India—New Delhi. | Technology and state—India—New Delhi. | Technology—Anthropological aspects—India—New Delhi. | Political science—Anthropological aspects—India—New Delhi.
Classification: LCC HD9685.I43 N4915 2017 (print) | LCC HD9685.I43 (ebook) | DDC 333.793/2095456—dc23
LC record available at https://lccn.loc.gov/2016043403

Cornell University Press strives to use environmentally responsible suppliers and materials to the fullest extent possible in the publishing of its books. Such materials include vegetable-based, low-VOC inks and acid-free papers that are recycled, totally chlorine-free, or partly composed of nonwood fibers. For further information, visit our website at www.cornellpress.cornell.edu.

Contents

Part III. Urban Transformations

ILLUSTRATIONS

PREFACE

When the lights went off across most of North India in August 2012, the international news organizations covering the event proclaimed it the world's biggest blackout—affecting the most people of any such event, ever (though they admitted that they included vast areas with sparse access to electricity in their assessment of how many people were affected). As a media event, this blackout provoked commentary from around the world. Experts on energy economics went on television news to explain the state of India's grid, and compare, invidiously, India's power production against Brazil's or Russia's (since these countries have comparable economies and territorial extent). I was watching this coverage and saw one expert say that a large-scale blackout could never happen in Brazil because the government had been investing hugely in its power grid, installing new transmission infrastructures, and generally supporting investment in the sector. I was surprised by this sanguine confidence in money and materials. Electricity grids are always more complicated than this.

Electricity grids require precise technical synchronization of generation and consumption, and they also demand a high degree of social coordination to handle loads that fluctuate with time of day, the season, and with ever-variable human behavior. Far from being a mere background, material condition of modern, urban life—operated through technical expertise and governmental know-how alone—the active management of the electricity grid is at once a cultural and a political practice in cities throughout the world, including in India. In New Delhi, to this day, planned power cuts help manage the grid (such cuts are called, in India, "load shedding," a technical phrase that is used in everyday talk about electric power). Running through practices of technical coordination, that is, are other processes of thought and judgment, which generate collective understandings of the power situation, cultivate expectations—even in the face of material limits—and raise expectations.

Blackouts and breakdowns, it is often said, uniquely provoke thought about the technological conditions in which we live, bringing into very clear focus the electric power in our homes and the multitude of institutions, devices, and relations through which it is delivered. This is certainly true of a massive event like the 2012 Indian blackout—almost a nationwide event, revealing the society-wide dependence on electrical power (even in places the grid does not yet reach, where the blackout was surely discussed too). But electrical power and the relations it fosters are, perhaps, more present to common consciousness than is allowed by this idea that people do not notice it until it is gone. Even when the energy is available and we can pay for it, electricity noticeably reshapes the space and time in which people live and creates new ways of making social distinctions and crafting judgments about life elsewhere. In their myriad, meaningful deployments, technological devices prod their users to comparative reflection about the life they are leading and the nature of life elsewhere—perhaps just on the other side of that boundary that separates day from night, or beyond the shadow-line of personal and technological history (how *did* people live then?). Electrification materializes and reshapes existing, moralized boundaries between, to take only two obvious examples, rural and urban, and the workday and nightlife. Electrical devices and the social capacities they unleash furnish the conditions—whether they are present or noticeably absent, on or off—for a certain world of sense and for judgments on the state of society.[1]

I was in India nine years before the 2012 blackout, in 2003, when the entire Eastern Seaboard and upper Midwest of the United States (and parts of Canada) went dark after a plant shutdown and a fire in Ohio put unmanageable strain on the whole regional transmission system. Then, the question asked by Indians and Americans alike had been, "How could this happen in the richest country in the world?" Some friends of mine in Delhi joked that America was becoming just like India, with its routine "load shedding"—a comparison that ironically inverted the more common equation, that with economic growth and increasing globalization India was becoming just like America. What this comparison (in either direction) indicated to me was that society-wide technological grids are formed through accretions of know-how, money, and material, but also meanings—they accrue metonymic associations and refer to a whole history of power and progress. In my initial forays in New Delhi, I found constant talk about "load shedding" and "power woes," and realized that people were making large claims about national history and political belonging as they discussed the terms of their connections and the conditions of their devices.

The grids and the connections of an electrical system attract comment in myriad ways and at multiple social moments—not just when they are new or break down, but as they routinely instrument social relations, too. When this happens they become not just the object of but also the occasion for wide-ranging considerations of power, freedom, inclusion, and collective history. This book tells stories selected from the political, legal, and technological history of electrification in the capital city of India and analyzes the meaning and force of electrical politics at distinct and often crucial moments of political, legal, and constitutional transformation in that country and in global regimes of power. It explores how processes of electrification and, subsequently, shifting patterns of regulation and distribution of electric power in New Delhi have focused governmental and popular concern on technology, shaped novel forms of governance and participation, and transformed both material standards and standards of judgment. It begins with colonial electrification and ends with neoliberal privatization, but neither the technical nor the political-legal processes here stand alone, disconnected from broader political events, legal currents, and cultural formations.

Under India's colonial government, even local electrification was enmeshed in empire-wide networks of capital and law. India's nationalist movement fought for a space of freedom from colonial economic control and cultural ascription, while also seeking to craft materially a new basis for interdependence and modern interconnection through regulation of basic infrastructures and industries like electric power. Finally, Delhi's experiments with electricity privatization over the past decade and a half have intervened in an urban grid that was already shaped by informal privatizations and political tinkering with citizen's connections, while instituting materially—in the form of new meters and upgraded connections—global ideas of good governance.

These broad processes of electrification are interesting, anthropologically, because they reveal cultural and political actors in the midst of action, questioning their belonging in time and place, and rethinking the very substance of society and its abstract yet real existence as the product of myriad interwoven energies. Debates over electrification, nationalization, and privatization—which are at once political and cultural—invoke rights and citizenship, theorize participation, and insist on certain criteria for legitimacy, as they seek to understand and to harness the modern sources of collective energy. These debates create and work with distinct "anthropological" understandings of their own, offering accounts of collectivity and commonality, crafting social distinctions, and operating with abstractions that run in parallel to more scholarly concepts of society, power, and order.

The analytic usefulness of *any* abstract account—scholarly or "native"—of what connects people into collectives and communities has long been contentious in anthropology and related fields. Contemporary advocates of renewed anthropological attention to materiality and to practical, everyday life often start by challenging the sociological reality of existing terms of analysis—such as nation, state, or society—and seek instead, in technical practices and in people's interactions with things, relations that are at once fundamental to everyday life as it is lived and that tangibly, if silently, connect phenomena across conceptual and geographical scales. Attention to such material practices, we are told, not only gets behind conventional social-science abstractions (such as "the state"), but also avoids culpably reproducing native mystifications, by revealing real connections that extend beyond purely ideological cultural or national boundaries.[2] Across a wide array of different intellectual orientations, anthropologists are often encouraged to take material connections, everyday interactions, and technical assemblages as the sociological ground beneath our ethnographic feet.[3]

By contrast, I seek to explore the intellectual and indeed often abstracting means political actors in New Delhi have deployed to understand and rework their material connections and their shared, social predicament, as well as the rituals, debates, and activism in which these have been expressed. These intellectual means range, in a rich array of encounters with electrification, from accounts of the past and of social change, to civilizational comparisons, to assertions of formal protocol and ritual status, to notions of private property and rights. I examine how these shifting understandings and social abstractions—of proprieties, properties, persons, community, and time—interact with technological installations, and limit or expand technical possibilities and political opportunities alike. Such an inquiry may also provide a basis for plugging the growing contemporary anthropological concern with infrastructures, networks, and the scope and reach of governmental powers within and beyond the state back into the discipline's classical and continuing traffic with ritual, magical power, and the formalities and forms of social ordering.[4]

In this book, it is the thoughts and practices of politicians, officials, citizens, and residents in New Delhi that matter. Before exploring these, I wish to outline briefly the contemporary relevance and resonance, for an anthropology of electricity, of the themes of technology, power, and ritual that were prominent in the work of the early twentieth-century anthropologist Arthur Maurice Hocart—a Sanskritist, anthropologist of ritual and caste, and colonial official (based for a significant part of his career in British Ceylon). Although Hocart's work is not addressed at length or in detail elsewhere in this book, starting with his ideas—which have something of the status of an alternative canon in anthropology—provides one way of introducing core concepts and methods that have guided this study of electrification in New Delhi. Not incidentally, Hocart also offered an interestingly idiosyncratic theory of how technology and ritual work together to produce and to legitimate time-and-space-spanning structures of state power, beyond their local materiality. Let us see what thus connecting the concerns of contemporary and classic anthropology might illuminate.

A. M. Hocart argued in his cross-cultural studies of ritual and government that even the most apparently mundane and technological tools for shaping the natural and social world were, first, ritual devices. In humanity's earliest forms of social life, he said, the most useful techniques and practices were those that concentrated thought and effort on the fundamental well-springs of life, centering power and affecting whole congeries of social relations. Moreover, he said that one could find the "old symbolism" of ancient rituals that connected people to each other, and to both mundane and transcendent spheres, woven into even the most modern of urban spaces. Hocart wrote, "The rationalizing historian can, of course, point . . . to the high vault of New York Central Station [*sic*] twinkling at night with electric stars, and ask triumphantly: 'Will you maintain that these architectural heavens are anything but a play of fancy, but pure art?'"[5]

Just because the ornaments and symbolic devices on otherwise useful spaces appear to be mere decoration, Hocart argued, does not mean that this is all that they are or do, even for self-consciously rational and utilitarian modern people. The abstract ideas of cosmic order and connection to centers of fundamental influence that are contained in the very idea of an astral constellation remain present in an electrical heavens—even if they may not be consciously grasped as such by those who pass beneath—and remain available for further, social uses. A railway terminal at the center of a great city is uniquely ennobled by the use of technology as ornament, and a whole cosmology is effectively woven into the material fabric of everyday, urban experience. The point may be clearer if we observe that such an elaborated edifice would seem out of place at a rural junction, its meanings misplaced and weirdly out of scale. This would jar with our sense not of utility but of social propriety (a distinction that makes the "rationalizing historian's" strictures beside the point). On this basis, we may begin to see why Hocart argued so strenuously that, for anthropology, the social and aesthetic aspects of a practice—or the ritual dimensions of a building—take priority over its narrowly utilitarian functions. Hocart's larger argument was that it would unnecessarily reduce our analytic purchase on the richness and thickness of human societies, our ability to work with the distinctions they make and to understand the distributions of power and agency they effect, to focus only on the material infrastructure that operates within such "useful" institutions, while ignoring their symbolic connections and coordinations of other, less tangible, powers. And he insisted that these symbolic and ritual aspects of useful technologies are, insofar as they are socially effective, eminently practical—they empower thought and give us purchase on the sources of social power.

On these terms, Hocart lamented what he called the "imperfection of our moral science," an impoverished modern knowledge that could not understand, however speculatively, the particular integration of experienced meaning with material powers.[6] Likewise commenting on our reduced modern intellectual equipment, the philosopher Isabelle Stengers has also challenged "the intolerant rule of abstractions that declare everything that escapes them frivolous, or insignificant, or sentimental."[7] In her philosophical work, Stengers draws on the metaphysical thought of A. N. Whitehead to describe the processes by which modern material life and its besetting powers are made meaningful and workable in the imagination and in collective experience. Hocart, not incidentally, was a contemporary of Whitehead's, and a real affinity exists, I think, between these two thinkers. Defining this affinity will help clarify what is at stake, anthropologically, in the modern examples with which I began, where there is some intellectual, comparative thought provoked by the society-wide organization of technological forces and artificially harnessed energies.

Working in very different grooves, Hocart and Whitehead both represent a kind of modern thought that takes processes of abstraction seriously and treats them as a generative source of social life, since these are what organize local energies and transform material connections into durable and distancing order and form. They both also recognized—as Stengers puts it in her commentary on Whitehead—that "abstractions may be transmitted either as 'living values,' values that incite curiosity, the appetite for contrasts, . . . [and] original responses to the situations of this world, or as 'dead values,' usually inciting compliant submission and the inhibition of what questions that compliance."[8] Both Hocart and Whitehead, then, in their different ways sought to challenge those intolerant abstractions that declared ritual frivolous, ornament savage, and (with a double rejection) non-Western "technological" thought both too meaningful and not adequately effective, and Western technology (by inversion) simply meaningless in its effectiveness. Both thinkers cherished insights that cut against radically materialist reductions and that highlighted instead how the world was made meaningful by multiple minds working in association with each other. They each sought to describe—one ethnologically, one philosophically—the acts of coordination and orientation (Hocart called them "rituals") that make of those minds a collective.

Hocart further argued that the prehistoric origins of modern-day "public services," including but not limited to urban technological infrastructures, were themselves to be found in ritual organizations for the production of life and prosperity. The instrumental use of power to effect change in the world, he posited, must first have been a ritual performance before it became mundane action.[9] The comparisons, as Hocart makes them, are fairly literal: like public services, ritual organizations—temples, festivals, and periodic rites—share their costs among all who participate in their benefits, and seek to produce and distribute (often unequally) good fortune, health, and continued collective life.[10] Although Hocart's point is exotic and perhaps even bizarre if taken as history or even as a speculative evolutionary tale about how we got from temple organizations to electrical infrastructures, it is not so easily dismissed if we recognize it as a comparative, ethnological observation about how people grasp social power and the effectiveness of their actions, in terms of the ritual technologies (the instruments of collective action) that they use to shape their time and place. The thought provoked by modern infrastructures and technologies of communication does seem to work upon deeply rooted understandings of social power itself. Electrical technologies and energetic infrastructures do produce power, and harness and convey that power to reach across distance and difference and connect and organize relations between humans and nature—much as ritual devices and magical forces have done in a wide variety of places and times.

Indeed, in his many comparisons of ethnological data with contemporary phenomena, Hocart tried to reveal modern technologies and forms of supposedly civilized statecraft as reiterations of ancient habits of mind and collective practice. As he noted, the once-magical "idea of action at a distance by the means of invisible forces has assumed increased importance once more in Europe since the discovery of electricity."[11] Meanwhile modern public utilities do, it seems, take on some of the organizing, life-promoting role that anthropologists have often found in ritual institutions and performances. What remains to be stressed is that for Hocart, the magical power of aesthetic coordination and collocation within ritual organizations was usually wielded on behalf of political power—it moralized state power and, bit by bit, garbed it in the borrowed aura of the good.

Part of what is at stake in contemporary controversies over electric power in Delhi is, quite directly, the morality of the state and the scope of its legitimate powers—whether and how state power should, on one view, foster the life of the community by providing the means, space, and time for work and leisure, or, by contrast, be limited in order to make room for private action. But something more diffuse and yet still decisively moral is also at issue in debates and discussions that have latched onto electricity as a means of progress and social transformation: wide distribution of electric power is taken to foster unity, integration, and political community itself, as it relieves the burdens of work and makes possible communication and connections across distance. It is not incidental that during the 2012 blackout, large swaths of the Indian population with no direct access to the electric grid that failed were, nevertheless, included in the count of those affected. More recently, electronic surveillance and a wholly transparent world of technologically mediated connections have promised a similar techno-moral reform of the political life of the whole nation, with the advent of new and more precise forms of technological regulation that also promise broader inclusion and participation. This latter is part of the rhetoric, at least, of the government's project of universal ID, or Aadhaar, led by tech entrepreneur Nandan Nilekani. Throughout the chapters that follow, electrical capacities and devices appear as mechanisms of work and connection but also as means of collective reflection upon the nature and extent of the social ties that make a nation. It is in this last respect especially that I treat electricity as a moral technology.[12]

This use of "moral" to refer to reflexive thought about, and action upon (and within), that which binds people together and connects disparate actors into a society or a community, may seem foreign to some—and indeed, this is not a common use in contemporary English. But moral has long had a sense, as in the phrase "moral imagination," in which it is more or less synonymous with "social" or "cultural"—with customs, habits, and expectations woven into the fabric of our common usage. Many such uses and customs may not ever be explicitly articulated, but nevertheless they are operative and can be discerned both materially and in action—like the associations complexly woven into a decorated train station or an electrical heavens. For some philosophers, accordingly, morality by no means forms the separate domain of a special science, but rather it is in play whenever the "nature or quality of our relationship to each other" is worked over in the course of politics, religion, play, or even technological installation.[13] Suffice to say, when we read about "moral science" (as with Hocart), or when the colonial state seeks to summarize in a government report the "moral and material progress of India," we are engaging with morality as some collective and workable arena of thought and action: the whole apparatus of understanding and judgment that makes the material world meaningful and accessible as something humanly shared—as a *social* world. On this understanding, society-wide grids and the technological devices they connect together are neither, only, relays of micrological interactions, nor just composed of physical associations of contiguity and force, nor simply mute material objects canalizing action. Such infrastructural technologies become part of whole worlds of sense and aspiration, and in this are always more than just material realities; they instrument and connect, shaping the material environments in which we move, to be sure, but also our less tangible, thought and unthought, relations to each other and to nature. As such, technologies are always deeply entangled with moral questions about how we should live, and subject to moral evaluation insofar as they operate to remake the common world. It is my hope that as an exploration of the confluence of the moral and the material in the environment of a great, technologically saturated, modern city, this account of electrification in New Delhi will thus resonate with tales of power and meaning from other times and places, where collective life is similarly instrumented, ornamented, and wired together.

ACKNOWLEDGMENTS

A friend once commented to me that the acknowledgments in an academic book provide the occasion for a kind of intellectual autobiography. Indeed, this book project has been part of my life for a very long time. But its roots are so entangled with interpersonal relations and social obligations of such depth and scope that autobiography alone cannot not suffice. Many people have helped along the way, and some readers who have commented on this text remain anonymous to me, while (writing being the affective struggle it is) my debt to others who have helped my research and thinking cannot be acknowledged in consciousness.

Still, some accounting will be worthwhile. To begin with, I am grateful to the institutions that have provided material support to this research. The Wenner-Gren Foundation generously funded my initial research in Delhi, while the Graduate School of Princeton University and the university's Fellowship of Woodrow Wilson Scholars provided consistent support at a very early stage in this work. A version of this manuscript was written during a year as a fellow at the Robert Penn Warren Center for the Humanities at Vanderbilt University in 2011–2012. The Ohio State University's Mershon Center supported research in Britain during which I snuck into archives to find out more about the Scottish connection to Delhi's electrification. At various times, the staff at the Delhi State Archives dealt patiently with a junior scholar, and the India Office Records at the British Library afforded excellent space and resources for archival research. In the last stage of this project, I was privileged to be asked to join the Department of Anthropology at Hunter College. I am grateful for the support the college has provided in preparation of this manuscript, and to my new colleagues for welcoming me into their conversations.

The moral support one receives in the academic life is, however, its true lifeblood. I am grateful to my graduate adviser at Princeton, John Borneman, whose guidance has shaped all my anthropological thinking—especially about ritual, power, and transformation. If his work is not cited in these pages, it is a default of presentation and might be taken as a sign of abiding influence. At Princeton, mentoring was very much a collective affair, and my work retains the impress of careful advising by Carol Greenhouse, João Biehl, and James Boon—whom I thank anew—and coursework and conversations with Abdellah Hammoudi, Isabelle Clark-Decès, Rena Lederman, Steven Kotkin, and Gyan Prakash. Formative intellectual influence, of course, produces its own kinds of transferential love and antagonism—I thank these teachers for withstanding both.

The faculty and students of the Department of Comparative Studies at Ohio State University formed the most collegial community of scholars I have known and expect to know—I cannot name each of them, but Nina Berman, Katey Borland, and Sabra Webber, as well as May Merganthaler, all pitched in with moral support at key moments, and I will always appreciate what I gained from working with them and with Eugene Holland, Barry Shank, Julia Watson, Tom Kasulis, and Philip Armstrong. Their faith in this project sustained it, and me, over several years of work. My friends in Columbus, most of all Isabella Winkler, Dana Renga, and Jennifer Siegel, remain across distance among the very best I have had.

From early drafts to final versions, I have benefited from the intellectual comradeship of Noelle Molé Liston, Erin Fitz-Henry, Talia Dan-Cohen, Simanti Dasgupta, Allison Fish, Jessica Zuchowski, Grégoire Mallard, and Eléonore Lépinard. My academic life has been enlivened and enriched at key moments by exchanges with Arudra Burra, Chris Garces, Parvis

Ghassem-Fachandi, Carrie Heitmeyer, Sarah Pinto, Eugene Raikhel, Martin Webb, and Natasha Zaretsky. Ritika Prasad and Andrea Ballestero both provided constructive and much-needed advice at a late stage in the writing, while Mark Harris's and Roy Dilley's teaching spurred me to become an anthropologist in the first place. Jim Lance, my editor at Cornell University Press, deserves a special thanks for his commitment to this project, and for finding the anonymous readers who provided such splendid comments on a manuscript that had not yet taken its final form. Kim Greenwell provided excellent editing services that helped get it there.

A small portion of the ethnographic material in chapter 5 was previously published, in a different form, in my article "Infrastructure and Interpretation: Meters, Dams, and State Imagination in Scotland and India," *American Ethnologist* 41 (2014): 457–72. I wish to thank the editor of *American Ethnologist*, Angelique Haugerud, for helping me shape that material for publication. Some of the research on which chapter 2 is based also provided the material for a differently argued article published in a volume edited by Roy Dilley and Thomas Kirsch, titled *Regimes of Ignorance* (New York: Berghahn, 2015). I appreciate their support and guidance.

Most of all, I am deeply grateful to the many people in Delhi and elsewhere who befriended me, talked to me, suffered my presence, gave me advice, or pointed me in a new direction in the course of this research. These debts extend from universities, to archives, to the neighborhoods where I sojourned, and beyond. A particular debt is also owed to my teachers and friends at the Landour Language School, whose generosity and patience provided not only instruction, but also renewal and spiritual strength at crucial moments in fieldwork. A project like this does not live without the patience, kindness, even the guardedness and discretion, of many interlocutors. I hope I have repaid in some measure their generosity with my own discretion.

None of these people bear any responsibility for what I have written. My parents, Jim and Helen, and my late grandmother, Irene, however, *were* responsible for my well-being for many years, and I share this book with them with love and gratitude. My siblings, Jimmy, Danny, and Sheilah, have remained close companions, mentors, and friends from childhood to adulthood, and they deserve much of the credit, as well, for who I am.

This book would not be a book at all were it not for David Petrain, to whom I owe the present richness of my moral imagination.

A Moral Technology

INTRODUCTION

Electricity Acts

Our technical practices are often as silent, as circumscribed, and as essential
as were the initiation rites of the past, but henceforth they are of a scientific
nature. It is in relation to these technical practices that historical
discourse is elaborated.

—MICHEL DE CERTEAU, "History: Science & Fiction," in *Heterologies*

In 1909, a young man from Delhi named Mohammed Asaf Ali set out
for London to study for the bar and prepare for a career in the Indian Civil
Service. On his journey, Asaf Ali stopped in Marseille, and he was immediately struck by the contrast between the busy, technological urbanity of
that imperial port city and the still relatively retiring urban life that he
was familiar with in Delhi. As he recounts in autobiographical notes written many years later, he was dazzled by the "charming boulevards and the
showily dressed shop windows," the "men and women moving together in
the streets, the uniform type of dress, the common use of mechanical ap-
pliances and gadgets . . . and the blaze of lights in the evening." "I said to
myself," he recalls, " 'So, Europe is all light, uniformity, mechanism, bustle, and freedom of women.' Delhi in those days was very different in all
these respects. I liked the change, and subconsciously began falling in step
with Europe."[1]

Delhi was, indeed, different from European industrial cities at this
time: the public, technological life of Asaf Ali's city featured only a small,

newly installed electric tram, and just a smattering of the private houses, mostly those occupied by British colonial officials, had electric lights.[2] Politically, while it was certainly a center of trade and commerce—and the largest city in its region—Delhi had been relegated for the previous fifty years to the status of a minor province within the British imperial structure. It had no important government offices, few of the municipal institutions associated with a major city, and scant career opportunities for an ambitious, educated young man who was seeking to reach beyond the cloistered, courtly world of Muslim Delhi in which he had been raised. Marseille, by contrast, was a prosperous, cosmopolitan, and technologically advanced city with electric streetlights and illuminated signs over shop windows featuring animated displays, while beyond it London beckoned as the center of the imperial government that ruled over India. One might well imagine that Asaf Ali felt some relief, and much excitement, as he entered this new world of "freedom and bustle." But "falling in step with Europe" was not all that followed from this arrival scene. "Later on," he writes in his memoirs, "after the flood of novelty and the dazzling effect of western lights had exacted their full tribute of admiration, I had time to think and compare the material and moral elements of this new world and of the old from which I came."[3]

This comparison between "old" and "new" is, at first sight, redolent of Orientalist contrasts between the cold, hard modernity of the technologically minded West and the rich culture of an affective, sensible, refined East. Asaf Ali also rehearses polarized oppositions between body and mind, sensuality and civilization, matter and morality, that were integral to the European colonial project and structured its definition of progress. The very phrase "material and moral" as Asaf Ali uses it was among the most clichéd elements in the colonial lexicon, underwriting a vision of progressive transformation in which civilization was synonymous with industry and advanced technology. The British government of India produced annual reports, "Statements Exhibiting the Moral and Material Progress and Condition of India," which provided an overview of the life of the colony and combined statistical measures of technological and industrial advance with reports on diverse social realities, including the status of women and the progress of "enlightened" mores among India's "communities." These reports were key props of a colonial ideology and practice that deferred freedom for colonized peoples until some unspecified level of civilization had been reached.[4]

For the young Asaf Ali, building on this colonial understanding of progress, the "common use of gadgets" in Europe revealed, materially as it were, moral and political differences between societies. Yet he declined to accept this technical contrast between old and new as a guide to India's future. Surely, he concluded, some accommodation could be sought between degrees of freedom and richness of culture, between the poles of the Urdu-speaking, courtly and refined society from which he came, and the world of material progress, gadgetry, and "uniformity" of status that he sought to join as a colonial lawyer. This politicized act of comparison is, he says, what set him on his career as a nationalist.[5] His recollections of his youthful journey and the meanings it had for him testify, that is, not only to a colonial knowledge-economy built from evolutionist notions of progress and invidious comparisons between technological standards; they also reveal Asaf Ali's embrace of a more widely shared modernist imaginary of community and social change, in which the manifold energies shaped by modern technologies and infrastructures, the things they move and the spaces they illuminate, could serve a moral renewal of collective life and herald the advent of political freedom.[6]

Asaf Ali wrote his autobiographical notes in the last years of his life, after a career as a nationalist politician and member of the modernist, progressive wing of the Indian National Congress (he died abroad, on diplomatic service, in 1954). Throughout his career, he served with distinction in his home city as a municipal councilor and as a patron of civic associations—among them, the Delhi Electricity Consumers Association—playing a role in Delhi's growth and expansion as a major Indian metropolis. The city became, in 1912, the capital of British India and gained the name New Delhi (by which it is still known internationally today), and thus the urban center to which Asaf Ali returned from London was already much transformed. Over time, through both colonial interventions and nationalist activism, New Delhi became a central place for a modernist politics of progress and planning, an urban home for the imperial and national governments of India, a site of a vigorous municipal nationalist politics, a laboratory for new constitutional arrangements, and, as a planned capital, a model of the power of centralized institutions to produce coordinated technological transformation. Electric lights and urban life, political changes and shifting populations, sparked intense debates over how, when, and why to install, expand, and govern new technological capacities and urban amenities. The whole conurbation, moreover, was a divided one, marked by the material and social differences between the "old" and "new" cities—staging a continued distinction between Indian culture and governmental modernity, and fueling further politicized comparisons, up to the present day.

This book is about electrification in New Delhi and how, at crucial junctures in India's history, this technological process has shaped and been shaped by stately rituals, political thought, and legal struggles. It explores the role technological transformation has played in wider debates over culture and community, political participation and national belonging across India's colonial and postcolonial history, by describing how discrete sets of contemporaries in imperial, postcolonial, and contemporary India—officials, politicians, citizens, activists, and even ethnographers—have, at different times, lived with and thought about electrical infrastructures, with all their power to conquer the night, redistribute work, produce invisible connections and interdependencies, and reshape space and time. While its examples are drawn from a single urban and national history of technological installation, regulation, and reform, together they address wider historical and anthropological questions about technology and freedom, state power and citizenship, inclusion and participation. It is important to say at the outset that this is neither an account of technological progress nor a book about how, technologically or socially, electricity became the necessity—and the need—that it remains today in urban India. Rather, through three case studies of electrification, urban transformation, and political mobilization in Delhi, it explores the entanglements between material projects and governmental, legal, and moral processes, and how these affect political consciousness and shape collective understandings of time, progress, and technology.

A Century of Urban Transformations

Over the past century and more, from early electrification to contemporary privatization, struggles over technological installations and the form of urban society have defined key moments in the history of Delhi and New Delhi; meanwhile these local, urban transformations have also been marked by political mobilizations and rituals of great pomp—often of national importance.

Electrification began in Delhi at the beginning of the twentieth century and was coincident with major changes in the city's political status; two great "durbars"—British imperial coronation rituals—held in the first decade of the twentieth century featured the installation of electric lights throughout Delhi, often obscuring older civic ornaments and signs of Indian sovereignty. These imperial rituals displayed the colonial command of modern technologies and also spurred Indian elites' efforts to install an equally spectacular electrical modernity for their own political and ritual purposes. All the while, the meanings and the machinery of these great rituals provided targets for Indian nationalists' attacks on both the justifications for imperialism and the technological practices of the colonial state. In turn, once these critiques had gained wide potency through the anti-technological and yet wholly modern thought and ritual practice of M. K. Gandhi, and India had won its independence, New Delhi hosted the new nation's constitutional assembly in a freshly renovated parliamentary hall.

In India's Constituent Assembly, fraught debates occurred under a brilliant constellation of new lights, about whether, and how, Delhi's citizens might control and manage their own public utilities. Delhi's representatives in this legislative and constitution-writing body insisted that it was the right of a big municipality to own and operate essential utility services like electricity and waterworks, and that a failure to provide for self-government on these terms would represent a moral rebuke to the aspirations of the independence movement as a whole.

Finally, in the twenty-first century, Delhi—now a city-state within India's federal structure—became a laboratory for global governmental experiments with power privatization and decentralization. Amid these reforms, new activist associations emerged that did not challenge privatization and the material improvements it brought, but rather organized and sued—unsuccessfully—on behalf of individual consumers' rights to control and manage their own electrical connections.

From imperial rituals, to constitutional debates, to civic meetings and legal cases, then, politicized comparisons between past and future, and here and elsewhere, that have also been moral evaluations of technological modernity, have cut across ongoing processes of urban growth, technological regulation, and state formation in New Delhi and India, especially as the power and freedoms conveyed by technological infrastructures were interpreted anew in order to challenge existing patterns of governmental authority, or to produce new ones.

Judgments about the meaning of electrification, its costs and benefits, and what goods it brings (or harms it imposes) animated the deliberations of colonial officials, structured postcolonial legal enactments and decisions, and infused India's modernist social science and intellectual culture more generally. In each period of India's modern history, such processes of reflection and judgment have shaped electrical grids and the connections they provide (to power and to other citizens), while also marking social boundaries and crafting new moral understandings of the common good. In addition to being an exploration of electrification and the regulation of technology in New Delhi, then, this book is also an inquiry into the political rituals and deliberative and communicative processes that have shaped modern political communities of varying scale and reach within the city and across India, ranging from neighborhood to nation, and from imperial to urban belonging.

Overall, New Delhi's urban specificity—as a capital city and a modern metropolis that capitalizes on all of India's rich history and diversity—enriches each of the case studies here and provides the basis for ethnographic comparisons that may help illuminate the wider relations that have been forged between technology and statecraft across our shared postcolonial modernity. At a minimum, to tell the tale of electrification in this city requires grappling with modern, urban technologies in all their material and meaningful dimensions: light, uniformity, mechanism, bustle, freedom, and splendor.

Moralizing Politics

The entanglement of technological infrastructures with political morality and images of the state has been a striking feature of India's politics from colonial to contemporary times.[7] As the political scientist Sunila Kale has recently discussed, the production and distribution of electricity served as an "illuminated sign of modernity [for] many Indian nationalists and political leaders," and the failures and inadequacies of India's electricity supply have long been "a cause for moral reproof."[8] These associations are evident even in quite recent political events. In 2013, a longtime anticorruption campaigner, Arvind Kejriwal, created a new political party called the Aam Aadmi Party (the "common man" party) and in an upset victory won the chief ministership in Delhi (he stepped down shortly thereafter, and then won the office again in 2015). In his campaign, he encouraged his followers to conduct a *bijli-pani satyagraha*, to engage in nonviolent protest against high electricity (*bijli*) and water (*pani*) bills. He modeled his efforts on the celebrated *satyagraha* campaigns of Mahatma Gandhi, though ironically Kejriwal's *satyagraha* was meant to produce more energy and more technological modernity, more cheaply, for Delhi's citizens—not exactly the moral import of Gandhi's campaigns.[9]

On the other side of the political spectrum, the current prime minister of India, Narendra Modi, leader of the Hindu nationalist Bharatiya Janata Party, has also risen to nationwide office by stressing his record of economic development as chief minister of the state of Gujarat; his campaign assurances included that he could deliver the long-promised goal of reliable power supply for India's citizens, or "24/7 power."

As always with such political trajectories, there is both tragedy and irony in the ascent of such a "strongman" to democratically legitimated power, with simultaneous promises of prosperity and broad-based participation on the one hand, and mobilization of exclusive and sometimes violent notions of belonging on the other. I once sat, in the summer of 2005, at an outside market in Ahmedabad (the capital of Gujarat) and was told that 24/7 power was *already* a reality in that city, because of Modi's political strength there and his no-nonsense management of the urban utility (incidentally, he had turned to a private company closely allied with him politically, Torrent Power, to supply power for Ahmedabad's public grid). What is ironic in this political achievement of broad-based electrification for (some of) Ahmedabad's citizens? Only this: I was told this while sitting at an outdoor market with some new acquaintances *during* a power outage. And this outage was clearly not an unusual occurrence, for when the local transformer exploded with a familiar cracking noise and the whole place went dark, employees of the many food stalls leapt into motion and with great alacrity got lanterns and candles going and lit charcoal to keep the market in business.

In this instance—as with Kejriwal's mobilization of *satyagraha* as a tool of protest against high bills and poor service in Delhi—an "urban amenity" (as utilities are often called in India) becomes a material site for the moralization of politics. Gatherings to talk about and work upon technological infrastructures become occasions for naming the actors who control the threads of collective life and for assessing their agency and efficacy; a public utility becomes the object of a struggle for the moral truth of political life, or else is associated with charisma of a leader who is uniquely able to provide developmental goods. In enticing and dismaying ways, Kejriwal and Modi *both* work over and recast problems with technological infrastructures as issues of India's public morality and collective life, and make these also into political issues in which state power is at stake. In so doing they repeat an older nationalist critique of colonial government and rehearse a modernist politics in which the technological devices of modernity became tokens—and tools—of political and social freedoms.

Critics have accused Kejriwal and the other anticorruption campaigners, as well as Modi, of profaning the sacred memory of Gandhianism with their highly instrumental deployments of Gandhian rhetoric, exploiting symbols of truth and freedom for partisan political purposes while reducing broad-scale political aspirations of solidarity and moral community to simple promises of development. It has been said that the heroic era of Indian politics is now truly over and that today's politicians are only pursuing pragmatic and—literally—*utilitarian* political ends, no longer concerned as they once were with the goal of a real collective freedom and moral renewal. The anthropologist Aradhana Sharma has pointed out, however, that the deployment of collective symbols and high moralizing in the course of mundanely material politics can be seen as very much in the mold set by Gandhi, who adroitly moved between the roles of *mahatma* and practical politician, discoursing about material need and leveling charges of mismanagement against the colonial power, while also engaging ascetic and religious repertoires, and producing a moral critique and claiming moral legitimacy, in his ritual performances.[10]

Analyzing such moments of technological change and moral concern requires an anthropological approach that can traverse the passage from material infrastructure to political imagination and back again. The Indian history recounted here—of political rituals and installations of technology, with both being used, together, to shape empire, nation, and city—is perhaps uniquely suited to such an anthropological understanding of technology and politics; and an anthropological approach may be uniquely suited to this history.

From Ethnographic to Historical Comparisons (and Back Again)

When I began fieldwork in 2005, the Indian capital was self-consciously a center of transformative political and cultural invention. Its government had already privatized electricity distribution in 2002—though the process was far from over—and it was pursuing further legal and technical experiments to reform all its public utilities, as part of nationwide efforts at liberalization (privatization of the water board, later halted, was a major goal).[11] Indeed, electricity privatization was just one element in an ongoing and interlinked politics of state reform and urban transformation—the newspapers trumpeted daily demolitions to eradicate illegal "encroachments" and improve public spaces, a new urban master plan was being vigorously promoted by civic bodies, and the Delhi Metro was in the last phases of its construction (the latter became a showpiece of Delhi's new urban sophistication and was, indeed, a radical improvement in the city's transportation infrastructure). At public meetings and in private conversations, middle-class citizens said that Delhi was on the verge of great things. They told me this, however, with an eye to the past, and the change that the present represented was measured against their sense of where their city had been and what political and technological promises had gone—and still might go—unfulfilled.

The statist legacy of planned control over public power that had characterized the postindependence period was going to be rolled back, I was told, and privatization would herald a new era of freedom and participation for all sectors of society. Deep and debilitating corruption in the public institutions that controlled Delhi's growth and prosperity was highlighted, time and again, as a problem of the present, and privatization offered one solution.[12] "A hundred years ago the utilities were private," a civil engineer told me at one point. He went on to observe, however, that the colonial infrastructures had been built without much Indian expertise or participation, making this into both an explanation of the present dilapidated condition of the grid and a lesson for its future: "If it [a new private utility system] is built without people's participation, if they don't know [how it works], how are they going to maintain it?"

Inspired by this and other conversations to do some historical research on Delhi's electricity system, how it had been built, and whom it had served, I found that, unsurprisingly, the narratives I was gathering were only partly supported by the historical record, but resonated with other claims that had been made and debates that had unspooled about Delhi's public utilities in similar contexts of change and opportunity in the past. This research helped me understand something yet more important than just how Delhi's infrastructures had been managed over time: namely, that fundamental political and even constitutional questions were recurrently raised when what was at stake was often just a technological matter, or an issue of cost, service, or metering. The moralized narratives I was being offered in the present, with private responsibility and imperatives of participation contrasted against public corruption, may not have worked as history—the electrification of Delhi under the imperial government had hardly been a beacon of business probity, while broad participatory governance was never a feature of state-run utilities—but they did cast a raking light on earlier periods, highlighting distinct and changing moralized narratives of past and future that were recurrently woven around electrical utilities in the city.

The standard history of Delhi's colonial growth and civic transformation, I found, mentioned electrification only twice, and only to say that the advent of this new urban technology in early twentieth-century Delhi was more influenced by the colonial state than by private enterprise: electricity arrived as a "side benefit" of the "Coronation Durbar" at which "the imperial state cast its gigantic shadow over the local."[13] The role of private companies was hardly mentioned there, though I later found out much more about the private company behind the tramway that ran in the old city throughout the first half of the twentieth century. That early avatar of private responsibility in Delhi sought a public monopoly from the colonial state, was undercapitalized, and suffered from bankruptcy (and shady dealing) among its contractors, which led to a growing consensus among colonial officials that they ought to "deprecate any attempt on the part of private capitalists to divert to their pockets the profits accruing from the running of a public utility concern."[14] The parallels between my conversations in the course of my fieldwork and the archival materials that I later read lay neither in similar technical or legal forms, nor in related patterns of ownership or investment, but rather in similar processes of judgment and politicized comparison.

In the cases studied in this book, we will see how the moralizing mission of British colonialism shaped both the purpose and the form of early technological installations in Delhi; witness India's postcolonial politicians debating how to combine technological progress with moral community; and explore how and why political debate in Delhi in the course of electricity privatization was seized by worries about corruption both in the state and in the neoliberal means by which technological infrastructures were now provided. Each case study here recounts debates over the public need for electricity in which its centrality to modern life was affirmed, for good or for ill, variously as a tool of government, a symbol of effective freedom, or a consumer good that still, however, needed to flow freely and cheaply. Explicitly moral and civilizational problems were posed by technological change in the colonial and imperial context: would it affect traditional customs? would it destroy communities? But long after these questions were rightly consigned to the proverbial dustbin of history, the moralization of the material means of modern life has continued to be a feature of political life in Delhi. In contemporary times, Delhi's citizens and even India's Supreme Court continue to assess the value and political morality of technical connections—whether they should be under local or central control, a state or an individual responsibility—and endlessly consider the means of managing them to produce beneficent effects.

Debates among officials, constitutional proprieties, and actions at law and in government are especially important to this account of electrification in Delhi. These are, and have been, the means by which people have grappled with material, technological systems that connected them to each other and to centers of power. Indeed, legal rights and problems of legitimate political action were the most salient topics of the contemporary debate over privatization and technological improvement in Delhi from which this research began. Legal questions about who owned the electricity meters housed in private homes and political ones about who should bear the costs of technical improvements were asked in multiple public forums. Meanwhile, activists explicitly framed their concerns by appeals to both law and propriety: Who was corruptly stealing power, and shifting the costs to others?

Such problems took the shape they did, an explanation on purely economic grounds might lead us to think, merely because this is how people talk about utilities—politically—in a context of unmet needs and in the face of social and technical pressures in a growing global city. Material shortage and technical inadequacy provoke legal and political struggle, or so goes the claim.[15] But such an account—in which the political is derived from the economic, as a reflex reaction to sheer necessity—can be effectively challenged, anthropologically, by situating this current politics in the context of the earlier case studies here. The terms on which power was distributed in imperial Delhi were different from what they are today—both politically and technologically—while the postindependence period is historically distinctive for any number of reasons. However, in all three cases, politicized comparisons between the technological present and conditions elsewhere or in the past (or, in other terms, between "modernity" and "tradition") gave substance and form to what were also, routinely, matters of economics or of utilitarian calculation. By putting these cases together one gains distinctive insights into the meaningful relations and associations that structure modern political life, as people forge connections between technological goods, belonging (or participation), and state power.

Grids and Groups

In many respects, this Indian history of technology, morality, and politics is a quite typical one: installation and expansion of novel technical capacities presented political questions and provoked experimentation and social comment in India as everywhere else. An anthropological approach to this history, however, focuses our attention neither on problems of the pace and timing of technological change—as a strictly technological history might—nor simply on the social or political effects of new capacities and organizations of power.[16] Rather, a broad anthropological approach of the sort that has guided my analyses can, I believe, bring into view a multidimensional process in which technological objects are taken up and their political meanings and social implications transformed in the course of social rituals or collective performances. The technologies that form and support urban modernity become part of whole cosmologies, as it were, and in turn are further shaped and reinforced by social practices of judgment and distinction that themselves deploy technological insignia, emblems, or devices. Grasping this process requires examining how such objects become meaningful, how they are interpreted as having collective effects, and how they are integrated (or not) into a larger understanding of time and collectivity—a diverse and often contested level of local knowledge about electrical power, what it is good for, what ends it serves, and whom it benefits. These local understandings are specific to a given conjuncture of events and cultural patterns, to be sure—as explored in each of the case studies here—but in the case of electrification and its meanings they also overlap with the conceptual repertoire of a wider modernist social science structured by oppositions between power and freedom, structure and agency, collective and individual, and by theories of progress that array such oppositions in a sequence of historical realizations. Let me briefly outline what, in particular, I find important about the historical interactions between modernist knowledge and local understandings of electrification, indicate the congruence between these historical thoughts and more recent critical accounts of energy economies and state power, and explain my cross-cutting perspective on the crafting of moral orders in and through technological politics.

The promise and the political potential of electrification were contested from the very beginning of its appearance in urban societies around the world. E. M. Forster famously imagined, in his short story "The Machine Stops" of 1903, a horror of etiolated bodies and hyper-regimented minds that would be produced by a wholly electrified society. M. K. Gandhi, as the spiritual leader of the Indian independence movement, turned such worries in a more thoroughly ethical and political direction, arriving at an "unusually intransigent rejection of modernity's material, technical, and political attractions."[17] Most importantly, Gandhi precisely identified the danger that electrification presented to autonomy and self-rule, writing in 1934 about then-popular programs of industrial and urban decentralization through rural electrification that "the power will come from one selected centre [and] that, in the end, would . . . place such limitless power in one human agency that I dread to think of it."[18] These early critics saw, clearly, that the social organization necessary to produce the energy that is distributed and widely consumed as electricity imposes a "rationale for the state and a grid of discipline."[19]

These words are from a 2008 essay by the ecologist and scientist Shiv Visvanathan, which explores the continued relevance of such modernist worries about electrification and technology to today's global economy and the scientific and technological basis on which it continues to expand. The consequences of ecological devastation, social domination, and uniformity of control that were feared by a Forster or a Gandhi remain tangible political perils in a networked, gridded, technologically integrated global society, as has been most recently pointed out by Dominic Boyer and his collaborators. They adapt the Foucauldian analytics of "biopolitics"—the comprehensive regulation of the life of whole populations by myriad and disaggregated institutions—to understand how contemporary polities and communities are integrally formed by energy infrastructures, with their autonomous logics. Boyer, using the term "grid" to articulate his account of modern energy, writes that "grid, then, is an apparatus subtly inclined to encourage demand, to expand itself, to solicit further dependency on its powers, which then grow in response. . . . It is not just a state instrument, in other words, a tool invented to accomplish a governmental agenda. Rather, grid must be understood as the organization of enabling power that allows any invention of statecraft to occur in the first place." Visvanathan and Boyer both point to alternative energy sources, low-carbon economies, and use of local resources as a "threat to grid" in this sense.[20]

The "physics of the steam engine" and ultimately the production, distribution, and consumption of electricity give rise, Visvanathan argues, to state power and a global "anthropology of modernity" characterized by incorporation in coordinated systems, hierarchy, control, and massive mobilization of means and matériel.[21] But there is another dimension to this very anthropology of modernity, this shaping of human thought and action by energy and its powers, one that comprises interaction and exchange across great distances, remarkable feats of power, magical devices, wonder, anticipation, and even joyous embrace of the new technological-cum-political world of bustle and freedom.[22] This "other dimension" by no means offers a simple "alternative" to the gridding, besetting, separating powers that Visvanathan, Boyer, and a host of other critics, have identified as features of modernity (and variously analyzed as the "iron cage," discipline, control, or biopolitics). Its processes neither contradict those other patterns of power, nor do they promise to revolutionize them. The two dimensions are neither opposed nor simply parallel, but interact historically to form specific patterns of social life, to incorporate but also to exclude, to regulate but also to empower.

That is, a focus on "grid" brings into view political processes of organization, control, and regulation, that have the effects, ultimately, of subordinating human creativity and vitality and producing life and populations as resources, to be expropriated in the service of power. Conventional central-station electrification does, indeed, require the coordination of myriad individual actions and situates the desires and energies of populations in subordinate relation to its own autotelic expansion (as Boyer indicates, and Gandhi earlier charged). These processes of "grid," importantly, are built up piece by piece to incorporate populations while also providing the basis for large-scale organizations that often expand far beyond state and community boundaries, hence escaping a critical—or anthropological—consciousness tuned to political and symbolic divisions. Grid operates both above and below the level at which any political consciousness that is not a fully abstracted expert knowledge, or a technical accounting of diffuse powers, exists and operates. This last point highlights why anthropologists have, with the recent turn to studying infrastructures and their materiality, embraced analyses that focus on dense interconnections and sites where technologies impinge upon bodies quite directly: exploring such material interconnections and interdependencies promises insights into micrological, material processes that at the same time have large-scale effects, which otherwise we could not grasp or understand.

But these processes of organization and control do not operate on their own. They must be worked, and reworked, through epistemological and evaluative processes that make specific claims to local relevance and mark out distinct connections across space and time—altering the apparently neutral topography of the grid in distinct ways. I wonder, then, if there is not some other lens we could apply, not in opposition to "grid" but rather to illuminate the ways in which what we might call (to anticipate the argument a bit) "processes of inclusion" cut across it.[23] This alternative lens would try to bring into focus the sometimes local and sometimes general and abstracting actions that interact with grid, potentially reshaping or distorting it, when a person seeks to grasp the power it distributes and simultaneously to assess the variable reach and effectiveness of her action. We might, to complete the borrowing that is already implicit here of Mary Douglas's well-known pair of terms, call this other dimension "group."

Now, when she introduced "grid-group" analysis, Douglas was not trying to make any cute puns on electricity. She was, rather, defining "a method of identifying cultural bias, of finding an array of beliefs locked together into relational patterns" in such a way that beliefs and processes of moral orientation became "part of the action" relevant to anthropological analysis (thus moving away from purely formal modeling of behavior on one axis alone, defined by a progression from constraint to freely expressed self-interest, or from "structure" to "agency").[24] Her formal model of grid and group was itself a two-by-two grid, with a vertical axis of regulation, hierarchy, and rule-bound behavior (low to high "grid") and a horizontal axis of degree of incorporation into a thick mesh of horizontal, affective, mutually responsive relations (weak to strong "group"). For Douglas, grid and group were not, however, opposed to each other, but represented dimensions along which disparate societies, or parts of societies, could be placed relationally; grid-group analysis was meant to reveal a variable and mutual interaction between, on the one hand, imposed constraints, established structures and institutions, and legal regulations, and beliefs, ideological patterns, and shared norms (which control and regulate, too), on the other.

Douglas's style of grid-group analysis cannot simply be applied to the details of the cases at stake here. Rather, what is important is Douglas's way of working through her examples (more than her systematic distinctions between them), which provides a means of incorporating into descriptions of cultural action both its meaningful, rich dimension, and the uneven distribution of social power and varying effectiveness of individual action in different social circumstances. She helps us understand how historical and scalar differences between types of social action might be produced and ratified through the orchestration of meaning and through forms of ritualized performance (not only through mute and material relations of force). Grid and group are not posed as historical alternatives or contradictory formations; rather, they are dimensions of any social circumstance. The approach, as Douglas developed it, was "capable of expansion and reduction between macro and micro scale social behaviour" and focused "on two modes of control—processes of inclusion and processes of classification."[25]

Grid and group are explicitly political concepts: they provide a way of gauging degrees of freedom and extensions of effective action, beyond the individual, through both cooperation and coercion. They can be thought of as a reconfiguration and interdigitation of Durkheim's famous heuristic contrast between mechanical solidarity—figured in his work as automatic constraint imposed upon identical units—and organic solidarity, with its note of mutual cooperation. The political thrust of Douglas's particular way of relating these two dimensions, with her attention to rich examples that exhibited *simultaneous* processes of "inclusion" and "classification," is an important part of her work's relevance today; this is especially so as political theories of the relation between the living, thinking, acting individual and the environing context of power, the network of cultural "grids" *and* communal bonds in which each lives, are perhaps no more well developed than when she wrote. As she put the political intention of her project, more by-the-by than I am treating it here but with a wonderful turn of phrase, "In his very negotiating work, each [person] is forcing culture down the throats of his fellow-men. . . . Their disputes are about standards and values, [for example] how town-life should be lived and justified, how country life should be supported, by this list of urban values and that list of rural ones."[26] On this account, values, beliefs, and judgments are as important as material organization or formal regulation for shaping political relations between persons; considerations of where one belongs and who one acts as (and whom one acts for), are the stuff of which political interactions and institutions are made, and an integral part of the way power is distributed.

For an anthropology of technological powers and political formations, taking "inclusion" as another dimension to the "anthropology of modernity" (which was initially defined, as by Shiv Viswanathan above, only by the harnessing and organization of forces and populations into one terrible machinery of power) means seeking out how historical understanding and collective action interact with besetting organizations of power, institutionalizations of authority, and biopolitical machineries that make one live in a certain way. This impels attention to the whole elaborate suite of political performances that can and do garland technologies (and infrastructures) with meaning, connect them up with a certain vision of sociality (positive or negative, hortatory or critical), grant them significance as attempts to forge solidarity, and distribute new possibilities of action (often, to be sure, only in the form of promise or aspiration).

The effort here, in sum, is to understand the broad historical and sociological effects of cultural, legal, and constitutional processes that work through infrastructures ("grid") by cutting, shaping, concentrating, or otherwise centering power, especially in rituals, moments of moral reflection, or decisive performances that provide a time and place for claims of status and belonging ("group"). Ceremonies, rituals, and procedure-rich legal and political occasions provide a site of reflection upon power and recursive interpretation of its processes. The effective operation of such processes of inclusion are not limited to democratic states or to popular forms of legitimation, or to rituals where a participant can recognize him- or herself, in an undistorted fashion, in the mirror of political legitimation. As I will argue, powerful calls to participation are also operative in imperial formations, even with their wholesale exclusions and narcissistic repertoires of justification.[27]

Finally, attention to "group" allows comparison between ceremonies and legal and formal events that we might not otherwise see as "rituals": for instance, a legislative debate, a courtroom drama, or a public meeting. What such "secular rituals" share is a heightened attention to order and form, a concern with protocol, custom, and symbol that emphasizes the cultural, collective address of their deployment of material, legal, and political powers. As Sally Falk Moore and Barbara Myerhoff have said, what defines secular rituals is the fact that "order is the dominant mode and is often quite exaggeratedly precise."[28] And they stress, as a methodological warrant for examining, together, such disparate events, "at our peril we ignore the form of [secular] ritual since that is always an essential carrier of its unique message."[29] The formal ordering and ritualized display of "gridded" classifications and besetting modern powers is part of what gives them what cosmological or broad cultural resonance they have—even when the meaning and significance of such events and the persuasiveness of their forms are contested or rejected. The powerful oppositions, separations, and regulations of grid cannot be explained by myriad, micrological interactions distributed across space and time alone—they must be given form and, more than form, human thickness, through their communication or performance (whether ritual, juridical, or political).

The Sequence of Forces and Historical Consciousness

Forms of belonging, claims of inclusion, and processes of community formation take shape, of course, in time and through myriad assessments of "the times"—historical consciousness and a sense of common fate and destiny are both essential to the formation of "group" as I have defined it here. This point is underscored by the ethnographic specificity of the processes and events in each of this book's case studies. The events, performances, and processes studied in this book are not organized into—because they do not fall into—one sequence of changing, evolving forces over time, nor do they add up to a conclusive, causal narrative. But put together and compared, they all reveal a pattern, a troping, of interpretations and actions in a recurring and recognizable form, one that may moreover be characteristic of a broader experience of technological and energetic change in the twentieth century. Namely, the events under study here reveal a particular questioning of history itself and of its collective meaning, which seems to be common to those contexts where and when new energies and new devices arrive and are self-consciously grasped in the flux of time to mark a new beginning and initiate a whole new social being.[30] This potent effect of technological change upon historical consciousness was acutely noted, at the very beginnings of the industrial, society-wide expansion of electrification, by the American essayist Henry Adams.

Adams, observing dynamos at work at the Paris Exposition Universelle in 1900, famously "began to feel the forty-foot dynamos as a moral force, much as the early Christians felt the Cross."[31] This experience of technological sublimity helped him grapple with a crisis of faith in historical progress that he had experienced in his efforts to understand his own era (Adams writes here in the third person): "Satisfied that the sequence of men led to nothing and that the sequence of their society could lead no further, while the mere sequence of time was artificial, and the sequence of thought was chaos, he turned at last to the sequence of force; and thus it happened that, after ten years' pursuit, he found himself lying in the Gallery of Machines at the Great Exposition of 1900, his historical neck broken by the sudden irruption of forces totally new."[32] The "sequence of forces," Adams stipulates, is the topic that modern minds turn to when their sense of man-made historical progression is broken by the irruptive appearance of energetic processes (or, simply, by technological novelties harnessing new and great powers). And yet, again and again, the promise of mastery is shattered by new forces. The question at the heart of Adams's crisis of faith, wrought by his encounter with electricity-generating technologies, is perhaps best captured in the critic R. P. Blackmur's gloss: "What were the forces that moved modern society, how did they hang together, *and how was a self-respecting mind to respond to them?*" (italics mine).[33]

In part because of the colonial context in which their thought was shaped, in part because of their hopes for the emergence of "the nation" as a self-conscious political entity, Indian nationalists—self-respecting minds responding to the conditions of their place and time—provide essential materials for revisiting the irruptive historical and cultural force of technology, and how it might be seized for common purposes, on the terms so lucidly laid out by Adams. In his *Discovery of India*, to take a well-known instance, Jawaharlal Nehru laid out the "sequence of forces" operating in India's recent past, writing that "from the age of steam we proceeded to that of electricity and now we are in an age of biotechnics and electronics. The age of social science looms ahead. . . . We seem to be on the verge of increasing enormously the power resources of humanity, and all manner of epoch-making discoveries hover over the near future." Even here, as he praised how science and technology were "building a new life for mankind," however, Nehru registered his own uncertainties about what all this meant, for India and politically. He wrote, "A doubt creeps into my mind. It is not lack of power that we suffer from but a misuse of the power we possess or not a proper application of it. Science gives power but remains impersonal, purposeless, and almost unconcerned with our application of the knowledge it puts at our disposal."[34]

Nehru's tone took a yet darker cast when he repeated much the same phrases a few years later in India's independent legislature. After the first use of atomic weapons in World War II, Nehru framed once again India's new independent destiny in energetic and epochal terms:

> We are on the verge I think of a tremendous development in some direction of the human race. Consider the past few hundred years of human history: the world developed a new source of power, that is steam . . . and the industrial age came in. India with all her many virtues did not develop that source of power. It became a backward country because of that. The steam

age and the industrial age were followed by the electrical age which gradually crept in, and most of us were hardly aware of the change. . . . Now we are facing the atomic age . . . and this is something infinitely more powerful than either steam or electricity.[35]

These pronouncements about India's place in a wider history of industrial change can, of course, be understood as an iteration of a conventional modernist account of the relation between energy and progress. As the science-studies scholar Itty Abraham has observed, in Nehru's speech a whole theory of social progress is elaborated in which this progress is "dependent on the exploitation of matter" and can be "demonstrated tangibly and 'objectively' through examination of a society's primary source of energy."[36] However, Nehru's comments do not pertain only to the exploitation of energy—they represent efforts at historical and social understanding that also do cultural and political work to summon a new collective subject.

That is, the marking of an epoch on energetic terms brings with it, especially in the context of a political tract or a parliamentary speech, an address to the consciousness of its auditors, and moreover their *consciousness of collectivity*. Nehru's rhetoric, complete with its energetic criterion for progress, also invokes a shared space and time for reflection, and calls forth a sense of a common predicament in the midst of ineluctable progress. It is not incidental, either, that Nehru celebrated the age of electricity as a prelude to the modern, governmental era of knowledge and power that he saw dawning (the "age of social science" that he mentions). Such rhetoric—which presents itself at first as mere description of history, culture, and technology, but also operates as a forceful collective address and even a kind of incantation—works in tandem with the technological harnessing of material forces to reshape shared understandings of belonging, time, and community.[37] As the anthropologist Michael Fischer has written, programmatically building on some thoughts from Karl Marx in the *Eighteenth Brumaire*, "Technoscientific structures and structural change can be explained neither by reduction to great man stories, nor by deterministic stories of power relations (whether class, colonial, imperial, or postcolonial). At issue is the creation of consciousness."[38]

Ultimately, whether in great public speeches or in more diffuse ritualized performances, politicians like Nehru try to offer a common vocabulary and shared understanding with which to grasp the mute stuff of material progress (indeed, to judge whether it is progress or not in the first place). In such communicative rituals, they deploy technological devices and energetic periodizations in public and collective ways, reshaping the very terms on which a wider people—the audience and addressee of rituals—can understand and shape their present.

Ritual and Power

Electrification in New Delhi—including the formation of institutions and the mobilization of far-flung resources as much as the actual technical work of installation and expansion—was regularly shaped by just such pronouncements and ritual performances, which were at once attempts at political action and greatness and efforts at interpretation of the present. These events were organized and led by political actors who also temporarily became ritual specialists, whether the colonial viceroy George Curzon; or Asaf Ali, Nehru, or Gandhi; or contemporary politicians and activists like Kejriwal. In each performance of power, the relevant actors pursued their own historically and culturally situated definition of the common good. But as wide public events, inviting participation and further interpretation, these performances could also shape a shared understanding of the present and work to define, describe, and integrate political and social orders. Take, once again, Nehru's pronouncements on the course of India's national history. Nehru does not merely make a historian's periodizations but also marks out the character of his modernity and specifies that a kind of collective ("personal" as opposed to impersonal) power is necessary to rise to its demands and reshape its very qualities for the common good. How far his performances were successful in actually shaping popular understandings is another question, not immediately answerable on the basis of his speeches alone—but that is not the main question pursued here.

Aside from the reception of these speeches and addresses, it is equally important (and in its way, also ethnographic) to ask how these performative pronouncements and rituals pose questions about their own historical character at the same time that they provide the occasion for new installations of technological devices. In these ritual and performative enactments of technological modernity and governmental reform, historical discourse is articulated through a technical criterion of progress. This is one basis for ethnographic comparison between them, despite their distinct (and thick) historical and social contexts. Whether in the contemporary political life of New Delhi or in encounters with electrification in Indian villages, one finds again and again "self-respecting minds" grappling with a sequence of forces and how these are shaping the very moral fabric of their society. At key moments throughout Delhi's modern history, political actors raise questions about the constitutional implications of a technologically saturated public sphere, while citizens work to develop specific, collective responses to the challenges posed by new developments in electrical technology and by the often unequal distribution of its energies.

Such events call for a mode of description and analysis that can grasp their myriad and powerful dimensions, extended in reflective thought and meaningful practice as much as in instrumental action and realized as order and form, and as aspiration and enchantment, as often as discipline or domination. One model of such a complex approach to these kinds of ethnographic events might be found—mutatis mutandis—in Clifford Geertz's study of "ritualized water management" and the simultaneous ceremonial centralization and material distribution of power in premodern Bali. Geertz wrote that attending to actors' own interpretations of time and belonging, and the social effects that flowed from the performances they stage, "restores our sense of the ordering force of display, regard, and drama."[39] Moreover, as he argued elsewhere, taking events that are at once decisive for the life of society and—because of that—"symbolical" as objects of anthropological investigation hardly involves sacrificing a concurrent understanding of history, power, or materiality. In Geertz's words, "If charisma [or ritual concentration of power] is a sign of involvement with the animating centers of society, and if such centers are cultural phenomena and thus historically constructed, investigations into the symbolics of power and into its nature are very similar endeavors. The easy distinction between the trappings of rule and its substance becomes less sharp, even less real; what counts is the manner in which, a bit like mass and energy, they are transformed into each other."[40]

Another, more philosophical approach that likewise illuminates the mutual relation and interaction between material ordering and ritual

organization, between pomp and power, with similar methodological consequences for the study of material networks of power and what they distribute, can be found in Giorgio Agamben's recent genealogy of "economy
and government" in the West (a genealogy that is worked out historically
in colonial and postcolonial states like India in highly particular ways).
"Liturgy and protocol," Agamben says, are "constitutive of sovereignty"
and remain inherent in even the most rationalized, governmental forms of
power. Let me put Agamben's unsummarizable exploration of ritual and
governmental order in this way: ritual incantations and solemn invocations,
however superfluous, other-worldly, and "merely" ceremonial they may
seem, are essential to the workings of power, even disaggregated, "gridded," and governmental power, because they project into glorious futurity
the ends of government and figure in the here and now an image of a final
perfection of perpetual rest and perfect communion. That latter image,
reproduced in ritual acclamations, forms the kernel of sovereign power
insofar as it properly belongs to the inoperativity of the human but instead
is used to legitimate exactions of labor, to produce wider "economic" or-
derings of force, and to sustain further impositions of sovereign power.[41]
That is, through formal ritual, political actors—following the scripts they
have inherited and tailoring them to the moment—put glory, protocol,
and doxa or law (the tools of sovereign power) in a central place, which
(Agamben stipulates) ought rather be left empty for the unrepresentable
"people." By rehearsing the forms, the garb, and the legality of sovereignty,
formal ritual action resolves (for a time) the constantly open, aporetic relation between the "economic" ordering of forces and the ends of political
power—and with this secures, with ceremonial or liturgical cement, the
capstone of the "great arch" of a properly *state* power. In Agamben's words,
"profane acclamations are not an ornament of political power, but found
and justify it."[42] In this endlessly repeated effort, ritual does more than
"legitimate" power—it *nourishes* it.[43] The chapters that follow show how
this reiterated work of founding—or, perhaps better, inauguration, since
what is ethnographically at stake in each case is the advent of a new power,
and a distinct set of expropriations and new moral orders—happens, in the
course of specific rituals and governmental projects.

Indeed, at the risk of seeming to make a crashingly literal equiva-
lence: it is just this aporetic relation between ritual order and economic

distribution, as Agamben defines it, that is recurrently at stake in the politics of electrification in Delhi, and which is resolved distinctly each time, in the course of building and legitimating different regimes with their distinct logics of organization and of besetting "grid." This aporetic relation is not to be found inside some technological device, simply there to be dissected or disassembled; rather, this relation is figuratively present and must be discerned and interpreted, time and again in locally problematic sets of (ritual) practices in which glory—transcendent ideas of state power, beneficence, and prosperity—and the material organization of an urban economy are hitched together (or undone). Although such relations between the material and moral are not *internal* to technologies, in their materiality, these relations are emblematized—and debated and transformed—in the form of meaningful devices. This latter emblematic aspect of technologies does not emerge from infrastructural materiality or the adoption and use of devices alone—it becomes salient only in the theoretical and political reflections and actions of the protagonists who operate and embed these relations, and through whom an art of government, and a claim of collective belonging and being, is launched, ever anew.

Politicians and other historical actors stand, in this history, as points of interpretation and as bearers of an institutional, ritualized, role-specific charisma, not (as they often try to present themselves) the self-sufficient creators of the forces they invoke nor in sole command of them.[44] But though this set of processes is not to be reduced to "great man" stories, still, acts of founding, high political rhetoric, and much-rehearsed tales of central places and powerful people all matter here. What is at stake is the crafting of a political sphere and the founding of power (in diverse ways and with disparate consequences for participation and freedom). As Hannah Arendt has written, such a political space—populated by things and persons—cannot be constituted out of pure materiality and anonymous forces alone. "Power preserves the public realm and the space of appearance, and as such it is also the lifeblood of the human artifice, which, unless it is the scene of action and speech, of the web and human affairs and relationships and the stories engendered by them, lacks its ultimate *raison d'être*. Without being talked about by men and without housing them, the world would not be a human artifice but a heap of unrelated things."[45]

At last, to update Geertz's stipulation about the compatibility of symbol and power (quoted above), the historical-ethnographic and interpretive approach to technology and power adopted here does not demand that we anthropologists focus only on meaning to the exclusion of life and social being—or pursue "interpretation" at the expense of "ontology." In the course of the ritual, legal, political processes that actually make modern technologies ubiquitous and integral elements in our societies, technological infrastructures are in fact realized as something other than just a medium for social relations, or only disciplining devices that apply micrological power to individual bodies. They become interior to and elements of the very social relations that compose our always-corporate bodies, and their meaningful presence is also part of any properly social ontology. To borrow some terms from the metaphysical thought of the philosopher A. N. Whitehead, through ritual appropriations and in the social and legal processes that make them socially and collectively workable, material technologies become "ingredients" in "actual occasions"—and it is as such actual occasions that collectives gain historical being.[46]

Extended Case Studies: Law, Ritual, Transformation

As I have said, this book is not intended to be either a descriptive ethnography or a linear, chronological history of technological change in India. The so-called progress of technology is less important here, as I have argued, than the generation and transformation of contrasts and comparisons that root collective meanings and mythic distinctions within the very materiality of infrastructures, and vice versa.[47] In public debates over an installation, a connection, or a device, symbolic values are forged, meaning is granted to material or technical things and relations, and a common understanding is worked into existence of "how we continuingly stand or have stood to others and to power."[48]

Instead of a sequential account of material transformations, accordingly, the course of electrification in Delhi is divided here into three discrete case studies—presented in chronological sequence, but each separate historically and substantively and featuring discontinuous events and processes. Each case is extended around a crucial moment of legal change that is also a point of transition in the historiography of the Indian state and a major ritual or political moment in Delhi.[49]

Legal enactments, processes of regulation, and juridical questions form one pole around which each case study is formed. In particular, the extended events that make up each section of this book are concentrated around the years in which various legislative bodies with India-wide competence passed the Indian Electricity Act of 1903, the Electricity (Supply) Act, 1948, and the Electricity Act of 2003 (as discussed in the chapter summary that follows). These "Electricity Acts" do not add up to a comprehensive legislative history, but they do represent important markers in the wider course of electricity law and regulation in India, as I explain in each section.[50]

The other historical and ethnographic pole of each extended case is formed by what Don Handelman has called "public events" (a term he elaborates as a more embracing, less religious-sounding alternative to "ritual"). As with the deliberately broad, comparative sweep of the concept of "secular rituals," public events in this sense encompass ritualized performances, popular celebrations, and political dramas. Handelman's definition stresses that such events must feature organized communication, wide address, and transformative participation. This definition allows analysis and comparison, on equivalent terms, of monarchical ceremonies, constitutional debates, and public meetings, while contrasting all of these, together, as potent political performances against rote, ideological spectacles and the formal logic of the law alike.[51] The two poles of each case are never entirely separate, however, and public event and legal or constitutional enactment often inform each other.

Three aspects of the joint work of technology and statecraft are revealed in each of these three cases. First, in each case there is some installation and expansion of new technological affordances—which "rationalize" urban experience, participation, and belonging, producing the "everyday" of urban, industrial modernity.[52] However, this is not an account of numbers of connections or megawatts of productive capacity installed; rather, each case examines local sightings of and encounters with electrical devices and processes of electrification to capture particularly salient political and ritual "appearances" of technological goods within Delhi. Importantly, these "appearances" are always, also, effective social distributions of power and participation, as devices are installed in some neighborhoods and not others, or as generators are temporarily set up in the course of a state ritual only to be removed later. The emphasis throughout is on moments of experience with, and conscious reflection upon, the presence and power of technology, and on the claims of solidarity that are made on the basis of these installations. Second, each case highlights collective efforts to grasp and rework technology and its distributions of power—efforts that "invent" a particular device or form of connection as a symbol of belonging and as a means of grasping collective historicity.[53] Third, each of these salient appearances is analyzed for its *constitutional effect* upon the wider distribution of rights and social participation in the city—and the possibility of altering that existing distribution. I mean "constitutional" in a "sociological sense," defined by Max Weber as "the modus of distribution of power which determines the possibility of regulating social action."[54] Such a collective distribution and allocation of powers is a socio-legal fact as well as a technological one, as scientific know-how, technological capacity, and legal and political possibilities are interwoven in relations of coproduction.[55]

In each case I have made an effort to stay close to the thoughts expressed by different actors and thus to track shifting judgments and countervailing opinions, rather than to prosecute one particular line of critique—the political meaning of technology shifts, that is, with historical conditions and the thought that distinct technological installations provoke. The diversity of opinions expressed over time about the relation between technology and culture is very much to the point. Through these case studies, I show the incessant, pressing necessity, in anthropological study and in everyday political life alike, of *meaningful, moral reflection*, especially when we are confronted—as ethnographers or as citizens—with infrastructures and institutions that appear to be mutely material, besetting, enframing, and expropriating.

Chapter Summary

The history of early electrification under the colonial regime in Delhi is from the outset a dual one involving both ritual and administrative aspects of the colonial state. The case study that makes up Part I, "Imperial

Installations," tells both stories together, examining the ritual theories
about imperial government held by leading colonial officers and the way
in which these officials governed technological change in Delhi. Chapter 1,
"The Machinery of Government," focuses on events, both regulatory and
regal, that unspooled around the crucial juncture of 1903. In that year, the
first comprehensive Indian Electricity Act, with effect across all of British India, was passed by the Imperial Legislative Council—just as the
Delhi municipality was in negotiations with a group of British businessmen to install an electric tram. The legal formalities imposed on state and
the private licensee by this act were minimal but were given meaning and
wide effect through an imperial ritual politics that ran from the minor bureaucratic offices of the municipality all the way to the center of imperial
power, and that reached a simultaneously ritual and technological climax
with the 1903 Coronation Durbar. That monarchical ceremony was organized by Lord Curzon as a rite of imperial permanence and beneficence.
It blazed forth with great technological elaboration but, paradoxically,
left no permanent traces in the material fabric of Delhi. Its significance
lies, rather, in its ideological patterning and temporary technological instrumentation of the imperial relation between India and the Crown—a
relation that shaped routine bureaucratic politics and Indian princely appropriations of deluxe technologies, and ultimately set the template for the
grander Imperial Durbar held in 1911, at which the imperial capital was
transferred to Delhi, and New Delhi was founded.

Chapter 2, "Ritual Center and Divided City," explores the rationale
for and historical effects of the 1911 durbar, showing how its ritual logics shaped the construction of New Delhi as an imperial capital and the
installation of electricity there as at once an ornament and tool of the imperial government. To close this section, finally, I consider how the simultaneously technologically minded, ritualistic, and paternalistic imperialist
ideology was met and overmatched, symbolically, by Gandhi's unique
combination of utopian vision and ritual performance. While the material instruments deployed in service of these political projects were vastly
different, and their ideologies opposed, imperial and Gandhian ritualism
can still be fruitfully compared if we focus on the techno-political idiom
that they shared. While the imperial state sought to produce a modern,
electrically illuminated and technologically sophisticated material realization of overawing sovereignty in New Delhi, knitting together its power

with cables and wires, Gandhi emphasized a humble handicraft practice, spinning, that literally produced *thread* to be woven into the new cloth of self-reliance.

The second case study, "National Grids," moves away from entirely official and "stately" rituals in order to explore more explicit political debates over the meaning and purpose of electrification, in the era of constitution and nation building after independence. Chapter 3, "The Lifeblood of the Nation," examines the work of realizing Indian independence in the Constituent Assembly and among planners and politicians. Within the urban fabric of Delhi and in India as a whole, the social and political realities that the colonial state left behind had to be remade once independence was won—the infrastructures of the modern Indian state and their provision of the conditions of a life in common had to be reformed nationally, which meant reworked *legally* to accord with new conditions of democracy and reinstrumented *symbolically* to mean freedom and not dependence. Accordingly, at the same time as it worked to write a new constitution for India, the Constituent Assembly debated and passed an Electricity (Supply) Act, in 1948. Moreover, in the closing months of the Constituent Assembly's existence, in 1949, the status of Delhi as a capital city was once again placed under scrutiny: specifically, debates occurred about how the city might rule itself and provide for its own technological and social growth and development. These debates restaged, at an urban scale, the core questions that confronted the new nation's government: how to achieve technological and political integration of the national territory; and how to improve the circulations and interconnections that made for a real, materially rich citizenship and true freedom.

Chapter 4, "Broadcast Mantras," further explores the political meaning of technological self-reliance in independent India and surveys contemporary ethnographers' descriptions of the uses and meaning of modern technologies in postindependence Indian society. India's modernist processes of national integration, planning, and technological and economic regulation have been seen as the decisive elements in a "passive revolution," a politically instrumental expropriation of local powers and democratic potentials that entrenched a corporatist, capitalist state elite at the centers of national power.[56] Through analysis of contemporary ethnographic texts that offered far-flung sightings of radio-broadcast religious rituals and of novel uses of technological devices across India, I offer a different perspective. This chapter focuses on recovering historically salient under-standings of infrastructure and electrification projects, in which they were seen to provide a new basis for local power as well as to offer the promise of integration into expansive national freedoms. The actual achievements of the postcolonial state, its real legacies of statism and expropriation, re-main open to critical evaluation. However, in this section as a whole I seek to capture still-undimmed hopes for deepened and strengthened forms of democratic participation, as they sounded together with questions about the social impact of technological rationalization in constitutional debates, legal enactments, and ethnographic descriptions.

The open, democratic, and solidarizing vision of technology that I find in the postindependence period contrasts sharply with the possessive and privative register of the discourse around electricity and its devices that became dominant under privatization and neoliberal regulation in con-temporary Delhi, which is the object of the closing case study, "Urban Transformations." The final electricity act considered here, the central government's Electricity Act of 2003, represents the legal centerpiece of this process and was partly based on Delhi's own earlier reform of its local electricity laws. The two principal achievements of the 2003 act were deregulating individual private production of power (for domestic and commercial purposes) and criminalizing theft of power. The latter was the more consequential legal reform, making "non-bailable offences" out of long-standing collective forms of appropriation of power from the still shared (but no longer "public") grid. Both of these changes also had well-nigh constitutional consequences—individualizing consumers, ab-stracting their consumption practices from the wider distribution grid, and changing the relevant locus and apparatus of state power in relation to electricity consumers.

This legislation, on its face, reveals a shift from state concern with the quality of the *collective* connection to the grid to a neoliberal rationality that rests on policing the behavior of the individual consumer. However, when these laws took force in Delhi, multiple affects and conceptions of citizenship were already bound up in the electrical infrastructure, ma-terially speaking. New devices attached to the old grid revealed existing interdependencies and reanimated them as bases for political mobiliza-tion, as I show in chapter 5, "The Life of Property." This chapter focuses ethnographically on the politics of meter replacement and the activism of

Delhi's middle-class Residents' Welfare Associations (RWAs), while exploring legal cases and jurisprudential interpretations that finally closed off certain avenues of potential political change. The politics of the RWAs was neither just a materialist, consumer-based politics, fixated on prices and regularization of connectivity, nor the direct outgrowth of neoliberal imperatives; instead, as they sought to understand and use new electronic meters installed by the privatized electricity companies, these groups fostered an active cultural politics that had the effect of recasting solidarity and interdependence at the neighborhood level. Further, as I discuss in chapter 6, "A Model Colony," supposedly superannuated political affects continued to be expressed in the new urban dispensation, in both a nostalgic register of regret for the promises of the past and as a new claim on power and participation. The nostalgia and regret of some citizens were part of the same affective reworking of the past that oriented and made powerful the jurisprudential interpretations that entrenched privatized corporate control over Delhi's infrastructures. Ironically, each actor in this affective, interpretive politics mobilized a different interpretation of the past and the present, but each made materially effective newly privative understandings of power and connection.

Technological Moralities

The political scientist and India scholar Philip Oldenburg long ago argued, on the basis of a study in Delhi in the late 1960s, that public utilities in Delhi "function well *because* of 'political interference'" (emphasis his).[57] With this claim, he aimed to challenge the normative distinction between bureaucratic administration and interest-driven politics and to highlight how important political connections were for the management and regular operation of technical services in a "developing" city like Delhi—and for the broad and general distribution of their benefits. With this argument, he, too, was making a politicized comparison, challenging the developmentalist "common sense" of his day that favored apolitical administration, by highlighting Delhi's existing political management of its grid and claiming it as a qualified success.

Perhaps technical fragmentation and bricolage, which are more characteristic of Delhi's privatized infrastructures today than comprehensive saturation by political power, can also ultimately produce something like the old "political" quasi-universality of access and connection in Delhi. This latter idea is put forth by Ravi Sundaram's contemporary analysis of what he calls Delhi's "pirate modernity," in which the technological conditions of modern urban life are constituted and made nearly universal through practices of reuse, recirculation, and creative bricolage.[58] Sometimes, his work reminds us, the practices of everyday life can provide resources for reaching across the gap between the fictive universality of state guarantees of citizenship and their always insufficient realization in material grids and networks of power; this is the hope of alter-globalization, of numerous small-scale mobilizations and attempts at new empowerment. Be that as it may, one can certainly find in Sundaram's phrase "pirate modernity" a nice rhetorical and political counterpoint to the *private* modernity that is today produced in the enclaves of the new middle classes in neoliberal, affluent, global Delhi.

It is the task of this book to explore, and recall, from within the private modernity of the present, those other political terms on which electricity and other public infrastructures have been installed, and their benefits distributed, within Delhi. Often, and still, the benefits and benefices of public utilities have been concentrated at ritual centers in the purlieus of New Delhi rather than being broadly distributed to support the "leveling up of common amenities" that was dreamed of in a different era, "bringing out circulations and relations that will bestow on the nine cities [of ancient and modern Delhi] close ties of a common identity."[59] What makes the difference between these eras—of imperial arrogation, political oversight, and private or pirate modernities—however, is not the supply of energy or the sophistication of the technology at any given moment. Rather, what matters in each case is the thought and action that is conceivable and enacted in each era—that is to say, the richness or poverty of the moral imagination.

Part I

IMPERIAL INSTALLATIONS

The key . . . to the emotionalism of imperialism is the transposition of
evangelicalism to wholly secular objects, or alternatively the translation
of secular objectives to a religious level. In a strict sense, its creed
was the consecration of force.

—ERIC STOKES, *The English Utilitarians and India*

1

The Machinery of Government

The student has to search beyond administrative institutions if he desires to understand the nature of a state; he must discover, if he can, the motive force which drives the machinery of government.

—Ishtiaq Husain Qureshi, *The Administration of the Sultanate of Delhi*

On New Year's Day 1903, the sun rose over a massive temporary city built around the commercial center of Delhi—a city of tents and lath-and-plaster pavilions, with new railway stations set in what had been a landscape of fields and villages, and new roadways lined with electric arc lamps laid out across the plain. All of this accommodated tens of thousands of people, officials from India's provinces and rulers of the "princely states" of British India, as well as distinguished visitors from across the empire, who had come to Delhi for a great ceremony organized by the viceroy, Lord Curzon, to celebrate the coronation of Edward VII as king and emperor of India.

That day, the viceroy presided over the centerpiece of an extended ceremonial that had begun a week before with festive arrivals of notables from across India—marked by gun salutes—and a parade with officials mounted on elephants, and which was to end a week later with a state ball. The technical arrangements for this two-week ceremonial were themselves a feat of some magnitude. The "Central Camps" comprised a spacious garden city of tents that lay north of the Ridge, a prominent geographical feature marking the northwestern boundary of Delhi, and they housed some thirteen thousand people (most of whom were "followers" or servants of the barely fourteen hundred Europeans and officials quartered there).[1] Still farther north, an amphitheater for the main ceremonies was constructed to hold nine thousand European and Indian elite spectators, with standing room for another three thousand inside the amphitheater and provision for thousands more on artificial hills around it.[2] Camps of varying opulence and scale for attendees from the so-called princely or native states were spread out at distances around Delhi calibrated to reflect putatively "Indian" distinctions (more like British discriminations) of status and honor. The outlying camps accommodated vast numbers of visitors from across India, as well as the "upwards of 39,500 men" from the Indian Army, both British and Indian, who marched and played in the ceremonies.[3] All together, these temporary durbar settlements formed an agglomeration "approximating to the area of London north of the Thames and west of Charing-cross," according to the London *Times* report of January 2, 1903. Within the main European encampments alone, "seven and a half miles of twelve-foot road and three and a quarter miles of sixteen-foot road" were laid out with more than a hundred arc lamps along them, while ninety-three hundred incandescent lightbulbs were supplied to light the tents.[4] The electricity for the Central Camps was provided by a power plant situated near the Viceregal Lodge (the sole permanent building constructed for the event). The lodge and the electricity station were nestled just under the Flagstaff Tower, a historic monument that stood atop the Ridge and provided the visual center of the whole plan.

This whole event was dubbed a "Coronation Durbar" by Curzon himself, who both devised the ceremony and presided over its principal events. He sought to associate a British coronation, one that had already happened in England the previous August, with India in some way, and drew on colonial understandings of the ritual repertoire of the Mughal kings who had once ruled in Delhi (and whom the British had dethroned and exiled forty-five years earlier). *Darbaar* or durbar is a Persian-Urdu word that means court or royal audience, and British government in India and beyond mimicked this form, especially in its relations with colonial nobles, princes, and "chiefs." By the end of the nineteenth century, and especially after the assumption of direct sovereignty over India by the British Crown, durbars—formalized and stately receptions—between British officials and "native" rulers were a key prop or medium of British imperialism in India (the form was even exported to eastern Africa). Such durbars, whether they involved a petty commissioner and a local landowner or a viceroy and a great ruler like the Nizam of Hyderabad, were also ritual occasions on which the "central symbol of the British state and focus of national loyalty—the Crown—was reworked . . . in relation to India and the rest of the empire," as the historian Bernard Cohn has noted.[5]

The Coronation Durbar, for which around a hundred members of India's colonial nobility traveled to Delhi, and which centrally featured the procession of these elites to give homage to the king-emperor—in the person of the viceroy who stood in as his representative and proxy in India—was at once an iteration of a common colonial ritual that displayed imperial hierarchy and the "Oriental" difference of the colony, and at the same time an exercise of governmental prowess in another, more modern and technological, idiom altogether.[6] As a rite of hierarchy and a display of governmental prowess, the 1903 Coronation Durbar in Delhi also had significant effects upon the legal and technical processes of electrification already under way in the city. While early electrification in Delhi was spurred by British entrepreneurs seeking to make their gain in the stock markets of London—and hence was very much a part of wider colonial processes of capital investment in urban infrastructure—the local course of technology and power in the city cannot be understood without grasping the ritual logics of colonial rule so massively displayed in the Coronation Durbar.

As they historically overlapped within the same actions and performances, imperial ritual and colonial electrification provide one shining example of the taught lines of filiation—both intellectual and practical—between the quasi-ethnological ritual theories and the legal and bureaucratic practices of the colonial state. Moreover, in the case of urban utilities, the strands of imperial practice combined to distribute power and regulate social action, spatially and temporally, with constitutional consequences for the imperial and the local state alike.

One might wish to treat such durbars as nothing more than an ideological gloss on, and a displaced performance of, colonial rationalities, as a ceremonial supplement to a weak and understaffed colonial armature of rule, or as "invented traditions" with important ideological effects but ultimately divorced from the material realities of power.[7] But the use of technological devices and quasi-utilitarian analyses to both organize and ornament the ritual and to justify its costs—and the continued force in later urban developments of these "modern" justifications of putatively "antique" rituals—point attention to the interdependent relation between colonialism's modern techniques and its ritual forms. As the anthropologist Andrew Apter has said of Lord Lugard's (closely related) ritualism in Nigeria, it "did not exist in ceremonial isolation but belonged to an elaborate cosmology and culture of rule expressed as much by the rational techniques of governmentality . . . as by political ritual."[8]

For his part, Curzon was both a powerful proponent of a conservative imperialism that would provide "permanent, protective" rule over civilizations in decline or otherwise inferior to Britain, and a theorist of sorts of the "special requirements of the orient" for "etiquette and dignity"; his account of the colonial mission in his speeches and writings is more or less stripped of the pedagogical, progressive, and liberal justifications that had been elaborated by an earlier generation of theorists of empire as "civilizing mission."[9] Indeed, despite his modernist rhetoric of "reform" and "efficiency"—deployed in his efforts to streamline the colonial bureaucracy—Curzon quite consciously sought to imagine and enact the empire as a vast and eternal architectonic structure.[10] Curzon's vision of empire took tangible form in Delhi with the staging and the technological instrumentation of his durbar.

In this chapter, I explore these imperial ritual logics of the 1903 durbar as they intertwined with technical and legal processes of electrification within Delhi itself, and in its local government. I start with the imperial ritual, and the arduous negotiations that occurred before the durbar over the question of a gift or boon for India to be announced in the king's name, continuing on to show how these ritual logics were enacted, materially, with the temporary transformation of Delhi into a durbar city. Second, I examine how these logics intersected with and ultimately affected the timing and extent of municipal electrification in Delhi in years immediately surrounding the 1903 durbar. In chapter 2, I turn to the longer-term patterning of space, time, and technology in Delhi with the later and grander Imperial Durbar of 1911, after which the imperial capital was moved from Calcutta, and New Delhi was built as a permanent ritual center.

Imperialism as it was practiced and performed in Delhi refracted modern governmentality and planning in ritual performances; its ritual logics in turn shadowed governmental techniques. Together, the routine and the ritual of imperial government defined a vision of empire and a constitutional order that was to shape the rest of Delhi's history.

Efficient and Dignified

By the time he decided to stage a grand durbar, Curzon himself had a great reputation as an organizer, and he often said that the watchwords of his administration were "reform" and "efficiency." But if these terms immediately appear to evoke classically liberal notions of economic rationality twinned with democratic legitimation, Curzon's imperialism gave them a distinctive centralizing and autocratic meaning. By his own account, in his (much praised) reforms of the colonial bureaucracy he aimed to produce the conditions for the decisive action associated with the efficacy of the mythic imperial administrator, the "man on the spot," uniquely informed about "local requirements" and better able to make judgments than any minutely educated and procedure-burdened bureaucrat. As Curzon said to an audience of civil servants, he held that the supervision and control of an administration—at whatever level—by "one man only" represented "the very best form of government" (adding, perhaps with some irony, though that would be uncharacteristic, "presuming the man to be competent").[11] Curzon's firm grasp of his power of decision and his certainty that he knew in each circumstance what was the right move—what we might call his conviction in his own convictions—are personal idiosyncrasies much invoked by his biographers to explain his failures as a leader and administrator as much as his successes.[12] But they also were based in a theory of imperial power as a permanent, aristocratic, disinterested, and all but global government by the British, a theory that he articulated throughout his career—building on the anti-democratic arguments of imperialist intellectuals and ideologues like Sir Henry Maine and Sir James Fitzjames Stephen. The 1903 durbar, insofar as it was organized, designed, and tightly controlled by Curzon himself, was at once a temporary and a temporalizing realization of that theory of power in both material and ritual form.

Curzon intended, from the start, for his durbar to be more than just a spectacle. He wanted this gathering together of India's princes—the "traditional" and now subordinate rulers of India—in order to celebrate their British sovereign's coronation to have a high and tangible significance for British rule in India, and for it to display the efficacy and might of British rule. He sought to further entrench British government in the borrowed, and much transformed, forms of Mughal court ritual—processions, audiences, and reciprocal gifts—and also to legitimate, by this association, the hierarchies of ceremonial precedence and recognition innovated by the colonial state, including orders of nobility bestowed by the British Crown and forms of recognition and honor like multiple gun salutes.[13] All this courtly show was meant, by Curzon, to cultivate the loyalty of the colonized by the association of the British Crown with the India's fantasied monarchical and despotic past. Meanwhile, for the consumption of the European audience, the durbar was deliberately staged in the shadow of key sites in the 1857 battle for Delhi, and aged loyal veterans of that war were paraded at the very height of the principal ceremony. Historical associations could be, and were, played both ways, indigenizing rule and memorializing conquest and violent subordination at the same time.

Alongside the trappings of glorious and beneficent kingship, there was a contrast with "despotism" to be marked, one that was just as important for legitimating British imperial rule: it was essential that the durbar display the principles of orderly justice and legal and economic self-limitation associated with the imperial ideology's modern, utilitarian, rationalizing side. In his initial announcement of his plans for the durbar, Curzon was careful to emphasize that this ceremony was also a material investment, the return on which would be loyalty. "I desire to assure the public," Curzon told his Legislative Council in September 1902, "that the proposed arrangements are being run on strictly business-like and economical lines." He promised, thus, that this would be an efficient investment in ritual—more than half the expenditure on supplies, infrastructure, and accommodations were to be recovered through public sale of the ritual equipment after the show was over, or reuse of technical goods by the colonial state.[14]

The apparent split that emerges here between two aspects of the state—efficient or utilitarian and ritual or affective—is neither unique to the Indian iteration of the British imperial state, nor was it invented by Curzon. A half-century earlier, at the very outset of Queen Victoria's increasing assumption of ceremonial, imperial titles and duties, the constitutional theorist Walter Bagehot had neatly divided the practical working of British government into two sides, the "efficient" and the "dignified," granting the first to high-minded administrators and politicians, and the second to the monarchy. Moreover, the two sides were complementary; the latter obscured behind a fiction of personal power the complex workings of administration and satisfied the needs of the people for "simple ideas" of how they were governed.[15] Within the government of India, however, these two functions were not as clearly divided as they appeared to be in democratic Britain: Curzon as governor-general (his official, statutory title) stood at the head of the administrative apparatus of colonial rule in India; but he also, as viceroy, personally represented the sovereign in India, and received in person all the honors due to a monarch. He was both a bureaucrat and a ritual functionary. His durbar perfectly expressed the interdependence of these two roles in the colonial government: a massive mobilization of resources and technique to organize and house representative delegations from across India, but all in service of a pure, unmediated moment of sovereignty in which imperial power and ritual hierarchy were to be tangibly realized.

In the same speech in which he promised to run the Coronation Durbar in a businesslike manner, Curzon spoke in almost mystical terms about the effect that such a collective celebration of the monarch might have. "The life and vigor of a nation," he said, "are summed up before the world in the person of its sovereign. He symbolizes its unity and speaks for it in the gate.... The political force and moral grandeur of the nation are indisputably increased by this form of cohesion, and both are raised in the estimation of the world by a demonstration of its reality."[16]

The two aspects of Curzon's theory of imperial government—as limited, efficient, but also paternalistically close and personal—come together in the ultimately failed centerpiece of his plans for the durbar: the so-called durbar boon, a gift from the British sovereign to "his" Indian peoples. Though no boon was ultimately offered in 1903, Curzon's plans for such a gesture—rejected by officials in London on grounds of constitutional propriety—did influence later durbars, and arguably it was the absent boon more than anything else that gave force to Curzon's own unique combination of the efficient and the dignified as the core principles of imperial rule.

The Boon Problem

When, in September 1902, Curzon announced his plans for the durbar to his Legislative Council (and conveyed them to London), speechifying about how it would be "much more than a mere official recognition of the fact that one monarch has died and another succeeded," he had yet to devise some way in which the elevating effect he hoped for from this ceremony might be achieved in practice.[17] In private correspondence with his political superiors in London, he eventually pressed this problem, arguing that in order to achieve its solidarizing, ritual aims, the durbar ought to include some personal action by the king-emperor—some great act of sovereign grace should be announced, he thought, some boon that would fix this event in the "native mind" as a moment of truly imperial legitimacy. At the same time, he knew that whatever was to be done to raise this durbar above the ordinary had to be achieved in the cost-effective and businesslike manner that he had already staked the legitimacy of the event on. What Curzon did not count on was that his idea of Indian statesmanship would run aground on British constitutional principles—the very principles of the separation between the "efficient" and the "dignified" parts of government that his practice of imperial government systematically blurred. While in Britain's colonial governments the proconsul, governor, or viceroy was vested with tremendous personal power—the "one man rule" celebrated by Curzon was not a fiction in Britain's colonies—the monarch in Britain had much stricter constitutional limitations on his action. This was the whole point of Bagehot's elaboration of a separate, dignified role for the Crown. This British constitutional precept was to become a problem as the Crown's dignity was ritually transferred to Delhi in the 1903 durbar.

Late in 1902, as the durbar loomed, Curzon finally wrote to London with his plan for a durbar boon, to be announced at the high point of the durbar ceremony just after the accession and coronation of Edward VII was proclaimed. By twinning the two announcements he hoped to associate, in the minds of Indians, the boon with the advent of the new sovereign, and hence to gain the ritual effect, the cultivation of loyalty, he had promised from this event. The boon was to be tax relief on a grand scale. Curzon noted that the budget situation was advantageous, and twenty million rupees (two crores) could be given back in the form of "remissions," particularly by rolling back the already symbolically important and much-hated salt tax (later to be famous as an iniquity of the colonial state, and the object of Gandhi's 1930 "march to Dandi"—a salt-making region—and his "salt *satyagraha*").[18] In official correspondence with the India Office on October 23, 1902, Curzon underlined the importance of making this change in connection with the durbar. "In our judgment, it would be an act of political unwisdom to let the opportunity slip, and to relegate these remissions to the dull routine of the annual budget," he wrote.[19] At the same time, he and his advisers were certainly aware that any such personal legislative action by the sovereign abrogated the constitutional role of the monarchy, within Great Britain at least. Indeed, in response to his proposal, the India Office raised the additional concern that such a boon might instill a dangerous expectation of extra-constitutional personal actions at all future coronations (at least in India). Curzon sought to assure the civil servants that any danger of such a constitutional misunderstanding, either in India or Britain, was remote.[20] Lord Hamilton, however, the secretary of state for India and, as the member of the cabinet responsible for Britain's India policy, Curzon's direct political superior, felt it *would* form a dangerous precedent to associate a change in policy with the personal powers of the king-emperor. Curzon's proposal to make tax relief a sovereign boon was, at the outset, vetoed.

This was not the end of the matter, however. Curzon was personally piqued by the whole philosophical rationale for barring his grand gesture, not just the legal terms in which it was expressed. The permanent undersecretary in the India Office wrote in a memo to Hamilton (that he then forwarded to Curzon) that "the whole theory of our system of government is that we raise in taxes not a penny more than what is required for the public needs of the country [and] there is therefore no margin out of which presents can be made by the Sovereign to the people. Either the present scale of taxation is needed for the good of the country or it is not," and either way, the decision was one of policy and not a matter that implicated the king.[21] In response to this presumptuous and prolix instruction of him in the constitutional principles of imperial government, Curzon sent an angry official telegram to Hamilton—on November 12—standing his ground on both the imperial form and the constitutional legitimacy of his plan. "Do not put me in [the] invidious position of holding a Durbar to which all India is looking forward with happy expectation, but which

I solemnly warn you that your decision, if adhered to, will convert into a disastrous failure." Curzon followed up with another communiqué a week later in which he complained at length. This was, he wrote, a matter of "Indian statesmanship," not of financial control or prudent government. Moreover, "the proposed prohibition [on any boon] will sacrifice one of the greatest chances ever presented to [the] British Government in India, will reduce [the] Durbar politically to a fiasco, and will convert [the] universal expectation of [the] Indian people into sullenness and even despair. I cannot assume [the] responsibility of being the instrument of so grave a blunder."[22]

In a remarkable move, Curzon then telegraphed the king's private secretary to enlist royal support for his idea of a durbar boon—clarifying that he merely wished to associate this policy with the occasion of the imperial coronation, not actually proclaim it as a personal act of the king himself. But the time for such constitutional niceties was over—Curzon's intransigence displeased his superiors in London, and it was clearly beyond the limits of his role even as viceroy to involve the king in intradepartmental squabbles. Curzon's friend and biographer Lord Zetland attributed to the dispute over the durbar boon a permanent deterioration of relations between Curzon and the India Office: "It was these things which gave to the summer and autumn of 1902 so profound and sinister a significance" for Curzon's later career.[23]

Ultimately, Curzon had to compromise and admit the principle of separation of policy from pronouncement, administrative decision from ritual action. To "rescue the Durbar from failure" he would announce, only in his own official capacity as governor-general and administrative head of the government of India, merely that he *hoped* to be able to cut taxes soon. Under strict instructions from London, he would refrain from making any specific promises at the durbar itself either in the name of the king or in his capacity as viceroy.[24] In this way, the empty form of the boon was maintained, but stripped of all actual content. No effective demonstration of statesmanship was to be allowed. Nevertheless, the durbar went ahead—to great British acclaim, but not without the suspicion in many minds, both British and Indian, that the whole thing was rather empty.[25]

To be clear, a tax cut announced at the durbar would not in all likelihood have materially changed the negative reviews the event received in the Indian press—as a wasteful show and an insult to modern Indians. It is fair to characterize as fanciful and certainly ungrounded—indeed, all but unmoored from reason—Curzon's quasi-ethnological notion that "in the East, there is nothing strange, but something familiar and even sacred, in the practice that brings sovereigns into communion with their people in a ceremony of public solemnity and rejoicing, after they have succeeded to their high estate."[26] But Curzon's imperial imagination, with its monarchical cast, its love of power, its consecration of sovereign decision, is clearly revealed in this minor bureaucratic spat. Moreover, to avoid a rush to judgment about Curzon's motives, it must be stressed that the usual colonial associations between such ceremonials and loyalty cheaply bought are notably absent in his rhetoric about the "political force and moral grandeur" that he hoped to foster through this massive durbar. An earlier viceroy, Lord Lytton, who staged a much smaller "Imperial Assemblage" in Delhi in 1877 to proclaim Victoria as empress of India, famously said that Indians are "easily affected by sentiment and susceptible to the influence of symbols to which facts inadequately correspond," and "the further East you go, the greater becomes the importance of a bit of bunting."[27] Curzon, by contrast, from the first sought to offer a tangible gift and a real boon to magnify the effect of the Coronation Durbar itself. I know of no remark from the hand of Curzon comparable to Lytton's invidious comparison between ceremonial accoutrements and real, hard political facts. Rather than pursuing imperial policy through pomp, Curzon rather tried to put tax policy and efficient governmental technique at the service of his imperial ritual (inverting the priority of Lytton's more utilitarian accounting, in which ritual served as an efficient tool of better government).

That said, Curzon did not think creatively, either, about what kinds of boons might be acceptable both to elite Indian opinion and to the British cabinet. Any significant political concessions, any changes in the structure of governance in India—of the sort that were already being demanded by politically active Indians—were unimaginable to Curzon. In a speech in 1904, Curzon stipulated this plainly, saying that "I do not think that the salvation of India is to be sought on the field of politics at this stage of her development; and it is not my idea of statesmanship to earn a cheap applause by offering so-called boons for which the country is not ready."[28] In his personal correspondence with officials in London he already underscored this, writing "I know of no extension of political privileges that may be safely made."[29] Neither taxes nor politics could be used to buttress the effect of the durbar, then—all that was left was to make of it a spectacle in which its grandeur was matched only by its fiscal and technical efficiency and its precisely calibrated organization of splendid effect.

Technological Signs of Sovereignty

To meet the modern imperatives of mass ceremonial, massive works were required to clear the plains north of Delhi—occupied by a number of villages and farms—and to transform the vacated land into a technological showpiece of empire.[30] Tents by the thousands were ordered and erected, and they were provided with all modern conveniences, including electric light. Miles of copper wire were used to power hundreds of streetlamps along temporary roads, and a steam-powered tram was constructed to connect the vast encampment with the site of the durbar celebrations farther north, and with the city to the south. The official historian proudly recounted the material statistics (some of which I quoted above) in the elegant volume produced to commemorate the event, writing that "fifty-four tons of bare copper wire and over twelve miles of insulated wire were used." This effort paid off. In the durbar encampments, "the current was never once interrupted."[31]

The colonial state thus took tangible, but temporary, form in the encampments arrayed around Delhi, in the snow-white tents arrayed orthogonally beneath the Ridge and in the command of modern apparatuses that the whole event displayed. At night, the "whole of the Road and Camp boundaries were outlined by groups of incandescent lamps."[32] Not incidentally, the effect, to European observers at least, was of an orderly, regulated, and in its way cozily natural terrain: "Nothing could be finer in its way than the view of the illuminated canvas city, seen at night from the neighboring eminence of the Ridge. The white of the tents lay like snow in the foreground, mapped out in symmetrical partitions by twinkling points of light."[33]

The electrical effulgence of the colonial state was most spectacular, however, as the assembled notables gathered for state dinners and balls in the refurbished and temporarily extended courts of the old Mughal

Figure 1. Viceroy's camp, Coronation Durbar, 1903. Copyright © The British Library Board (India Office Records Photo 430/79 116).

palace—the Red Fort or Lal Qila. Situated in the heart of Delhi itself, the marble and stonework halls of the Mughal emperors were refurbished for the occasion, and in fact enclosed and extended with temporary imitations of their original architecture so as to be suited for the large gatherings that were planned for the durbar. The work of Curzon's military supply officers (personally directed by him) turned "the ancient seat of Shah Jehan and Aurangzeb" into a fantastic, technological "fairy land." The Diwan-i-Am audience hall, once the site of the Mughal emperor's own *darbaar*, was taken over for the state ball and expanded to thrice its size, and was transformed, the correspondent for *Blackwood's Edinburgh Magazine* wrote, by the "blaze of numberless electric lights and the glitter of diamonds." The official history is yet more specific: "The electric light arranged in the . . . coffered panels in the ceiling, and [in] clusters depending from the center, threw a soft radiance on arch and pillar. The great white marble throne with its curving baldacchino stood out from the dark background with an almost dazzling brilliance; while a rich glow, reflected from the crimson columns and roof, suffused the remainder of the gigantic hall."[34]

Figure 2. The Diwan-i-Am prepared for the state ball at the durbar, 1903.
Copyright © The British Library Board (India Office Records Photo 430/11 13).

Meanwhile, the Diwan-i-Khas, a smaller and more intimate reception pavilion, served as a state dining room. There, the celebrated Persian
verse "If there be a paradise on earth, it is this, it is this, it is this," which
was spelled out in inlaid marble upon the cornice, had been obscured by
more modern signs of sovereign pleasure. The *Blackwood's* correspondent registered the change, writing, "We ourself [*sic*] confess to a feeling
of relief when we found that the arrangements for the electrical installation had necessitated the temporary covering up with a false cornice of
that breathless inscription."[35] It is hard to decide if the terrible power this
writer feared was more present in the ancient Mughal inscription or in the
glittering display of technological prowess, which profaned and—to his
relief—obscured it.

For all the conventional recourse to images of Oriental magnificence,
variety, and inscrutability, there is an undeniable element of modern awe
and enchantment in these descriptions of the durbar and its purposes. That
is, the European visitors to the Coronation Durbar were equally enticed
by the Oriental pageant and the modern stage on which it was placed.
Even when events occurred in full sunlight, as with the state entry in

Chandni Chowk—with Lord and Lady Curzon carried on an elephant and officials and princes in resplendent carriages followed by processions of retainers—observers reached for metaphors from their stock of theatrical experiences, and spoke of the parade as if it had occurred in the irreal and artful world of stage illusions and electrically projected fantasies. Mortimer Menpes, a popular painter who released a book of sentimental

Figure 3. Lord and Lady Curzon on the state elephant.
Copyright © The British Library Board (India Office Records Photo 430/78 23).

sketches of scenes of "native life" witnessed at the durbar, compared the state entry to one of the entertainments offered by the "electric Salome" Loïe Fuller, an American dancer who produced electrically illuminated (and erotic) stage illusions. Menpes wrote: "Slowly, the blazing procession came into view, unwinding itself and presenting a wholly new scheme of colour every moment. It was never the same. It was as hopeless to follow each effect in detail as to catch every flash in the swimming skirts of Miss Loïe Fuller. Each instant produced its own glory."[36] An apparently composite image produced to convey the splendor of the state entry shows Lord and Lady Curzon in a howdah atop the state elephant (which they did in fact ride into Delhi for the state entry), with bearers and, just visible in the background, telegraph wires providing a touch of sophisticated modernity.

The durbar city was, then, an instantiation of an almost instantaneous modernity, a massive demonstration of British technical efficiency and organizational skill, which also served as the framework for making the exotic colonial other close, allowing an intimate grasp of otherness—up to and including the mimicry of "Oriental" ritual forms by European officials. Indeed, Curzon was lauded (and sometimes mocked) by his contemporaries for possessing (as his biographer David Gilmour puts it) the "invaluable but very un-British gift of being able to take himself seriously at such a pageant."[37] Besides this supposed gift of Curzon's imperial character, it is equally important to note how deeply the meaning of this pageantry depended on the deployment of modern, efficient means and up-to-date technologies to achieve putatively Oriental, that is to say nonmodern and exotic, excesses of sovereign display and beneficence.

Opposition by Oriental Excess

The Coronation Durbar, as a stage for the motivation and ratification of British supremacy, demanded "traditional" displays by Indian princes, too, as members of the quasi-aristocratic structure of British rule in India—demands that the princes often met with highly sophisticated navigation of protocol and custom, but also through adaptation and adoption of the very material signifiers of colonial difference, measuring up to the demands of Orientalizing ceremony with great technological elaboration of their own. An appropriative counter-symbolism and display of an avowedly Indian technological splendor was elaborated in the encampments built and funded by the rulers of India's myriad "princely" states—at distances from the central durbar grounds precisely calibrated to reflect distinctions of status and gradations in the "native" hierarchy.

In these encampments, built and staffed at the expense of the individual states, opportunities to outdo the colonial state in both splendor and effective modernity were available for those with the funds and desire to do so. Particularly the large and wealthy states seized this opportunity. The camp of the Kashmiri ruler impressed with lights emitting a total of 120,000 candlepower; the maharaja of Patiala had, guarding his camp, "two colossal statues of knights in armor, bearing electric lights."[38] Meanwhile, Austin Cook, a young English engineer in the employ of Bhopal state (who stayed in the suite of the maharani of Bhopal and not in the Central Camps for European government officials, and thus had a better view of the diversity of the installations laid on across the durbar grounds), wrote that the "best [camp] of all" was that of the Gaekwad of Baroda.[39] This naïve and impressed description, apparently written in a letter (only a typescript survives in the archive), is replete with tropes of civilizational difference but also of "wonderful engineering," and is worth quoting at length for the impression it gives of both civilizational difference and technological prowess in this "native" encampment:

> [The Gaekwad] determined to build a palace for himself [at the durbar] which should eclipse all the others, and at his State in Baroda he had an elegant bungalow of teakwood built, and from there he had it transported and set down in Delhi. On the center of the bungalow was a huge dome 50 feet high from which projected a gilt spire, and at the end of it was a splendid display of electric light. . . . In front of it his staff was accommodated in tents and behind was the maharani's abode. She had to be removed from the vulgar gaze as she is a victim of that terrible purdah system! She and her ladies in waiting occupied tents, enclosed within an eight-feet-high fence of wickerwork, draped on the inner side with green silk. The furniture and drapings of those tents were gorgeous, being all of the richest Indian silk, and as one went from room to room one could almost imagine oneself in a small palace of fairy land. The floors were covered with carpets of wondrous thick pile. The electric fittings were splendid and in every room were

several statuettes holding up their blaze of light. In the center of the lawn, in front of the bungalow, was a large fountain which was illuminated by powerful electric lights of ever-changing colour. At the side was the reception "shamiana"—a large tent—and in front of it was an archway bearing the word "welcome" and at night this was outlined by electric light. The engineering connected with the electric light was very wonderful, I believe, and it was indeed an exquisite sight to see that whole camp illuminated. Fairy land itself surely could not be more lovely!

Given the constraints on the Gaekwad, on this official occasion, to present his power and authority within the terms laid down by imperial and Orientalizing notions of customary kingship, it is no surprise that his teakwood palace at the 1903 Coronation Durbar outdid in style and opulence the official encampments. However, on this account, it also outdid them in splendid modernity, presenting the Gaekwad as, quite literally, an "enlightened" ruler. The official historian Wheeler, less ingenuous than Cook, writes, "the camp of His Highness . . . The Gaekwar [*sic*] of Baroda, . . . replete with every modern luxury, at once suggested that the Chief who presided over it must be a man of great enlightenment [and] such indeed is the case."[40]

Gayatri Spivak has commented that the hereditary heads of the native states of India "came to feel [their] 'royal'-ness rather more strongly under colonial influence, writing [their] accoutrements on a European model, or even perhaps a European conception of a 'native' King."[41] For Spivak, such displays of royal insignia and stately splendor relate "catachrestically" to European aristocratic formations, installing in distorted terms this alien idiom of rule within Indian political relations. This capture of "native" kingship by the signs of imperial suzerainty is an important reality of the displays of the princely states. But as Andrew Apter notes in his study of British imperial rituals of culture in Africa, the appropriation of European titles, honors, and material goods into the ritual performances of African kingship was not only the reproduction of an imperial discursive edifice obscuring "real" native difference, but also represented the "harness[ing] of colonial forms of culture and value" to serve African elites' projects of government.[42] The princely states' encampments at the Coronation Durbar might, on these terms, seem less like a hybrid or mimicked modernity and more like a powerful appropriation of the colonial state's claims on modernity through a creative reworking of its Orientalizing sumptuary obligations and parallel investments in modern ritual technologies.

Indeed, in Cook's awed descriptions of it, the Gaekwad's display of electrical magnificence seems to exceed any totalizing inscription of "royal" formulae upon Indian kingships—a gilded, illuminated spire fifty feet high! An electric "welcome" sign! The Gaekwad, along with the other representatives of the wealthy, major princely states, not only displayed the licensed hybridities of quasi-feudal, aristocratic, colonial kingliness but drew also on the symbolic power and magical force of modern technological display, demonstrating a greater grasp of the surplus power remaining in the colonial contradictions between progressive, technological government and permanent, protective, and distant sovereignty than perhaps the British themselves ever managed.[43] We can catch, in Cook's description of this encampment, the glimmerings of an Indian rewriting and rewiring (if you will) of imperial forms of state—only just visible behind Cook's nervous oscillation between Orientalist tropes of decadence ("that terrible purdah system!"), and his awe at the electrical display ("the engineering . . . was very wonderful").

Taken whole, in the long and conflict-ridden frame of a cultural encounter, it is neither straightforward nor perhaps necessary to specify whose cultural "property," whose idea of progress and modernity, or tradition and "native" custom, was involved on each side of this pageantry. Curzon's adoption of "Indo-Saracenic" architectural motifs for the durbar amphitheater is no more "Indian" than the electrification of the Gaekwad's palace is "English." Nor are these mobilizations of goods and technique simply modern government in traditional dress, another chapter in the "invention of tradition," or the factitious use of antique forms of status in the service of the reproduction of an imperializing capitalism ("capitalism in feudal-aristocratic drag," as Benedict Anderson has called colonialism).[44] The "feudal-aristocratic" trappings actually represented a serious concern and object of governmental technique, shaping the very rationalities of the imperial government. Meanwhile, the contradictions—and crisscrossing exchanges—between utilitarian, rational legitimation and ritualized concern for Oriental difference were powerful within local governmental debates over the pace and propriety of urban electrification, too.

Bureaucratic Delays and Personal Assurances

At the beginning of the twentieth century, Delhi was already growing by leaps and bounds and was the largest commercial center in its region. The city was throwing out suburban developments, while the municipality was installing improved sanitary and health infrastructures. However, the imperial installations for the durbar, including massive projects of drainage along the banks of the Jamuna, and all the roads and wires laid out for the encampments around the city, barely touched the historic and vibrant center of Delhi. According to Narayani Gupta, the leading historian of the city in the late nineteenth century, the Coronation Durbar was "one of those extravagant occasions when the Imperial Government cast its gigantic shadow over the local, [and] a city momentarily came into existence adjacent to the living organism of Delhi."[45]

Gupta also observes that electricity "came to Delhi" as a "side-benefit" of this durbar. On detailed inspection of the governmental debates over electrification and its utility to a "whilom capital" like Delhi (in the British commissioner's words), however, it becomes clear that the effects of the durbar, with its ritual logics and technological instrumentation, were more pervasive than this: it spurred the municipality to take action on electrification and granted it arguments for the modernization of its streetlights, while also expropriating local powers and infusing this municipal politics with India-wide and empire-wide legal concerns and cultural proprieties.

More than a year before Curzon initially proposed a durbar to his Legislative Council, a syndicate of British businessmen, led by one Thomas Wilson, approached the Delhi municipal committee with a proposal to develop an electric tramway and municipal supply of electric power, "for domestic and manufacturing purposes in the city and suburbs of Delhi."[46] This syndicate was just that—a loose and speculative venture of private businessmen—although they presented themselves as having investments and connections in multiple engineering and technical enterprises, including Killick Nixon & Co., "concessionaries for several important Indian railways," and the Bombay Tramways company, "the most successful tramway company in India."[47] In fact, the Bombay (horse-drawn) tramway was being acquired by that municipality as its license expired, and the men with capital invested in that concern sought frontiers of investment the new field of electric traction. Delhi and Lahore (where the same syndicate was also seeking a concession) presented sites for new business opportunities. For his part, Thomas Wilson was the proprietor of the Bombay engineering firm John Fleming & Company, while also presenting himself as the principal of an eponymous firm based in London.[48] His multiple corporate identities—his letterhead included addresses for both John Fleming & Co., Bombay, and Wilson & Co., London—no doubt helped buttress his appearance of business solidity to the Delhi municipality. In his initial approach to the municipality, Wilson laid out the business plan of the proposed company and requested monopoly rights from the city to "supply electric power for lighting, punkahs [ceiling fans], and other manufacturing or domestic purposes to all public and private buildings."

The municipal committee to which Wilson wrote in 1901, promising to aid the "industrial development of the city," was a primarily consultative body that helped to provide some measure of Indian "participation" in the administration of the city, but whose real powers were strictly limited. It was made up of three local notables with titles bestowed by the British, two merchants, four bankers, one lawyer, and a smattering of other professionals.[49] The committee was presided over by the British deputy commissioner, who held an absolute right of veto and the privilege of conveying decisions of the committee to higher levels of government for action; it had no executive authority of its own, nor power of the purse. While municipal reforms in the later nineteenth century had promised a greater role for such local governmental bodies, their responsibilities were ultimately limited to street lighting, night-soil removal, and other basic activities, with little discretion for capital improvements such as electrification.[50] Most importantly, although Delhi in 1901 was a rapidly growing commercial city and was larger in population than Lahore (though not in European population), it was neither an autonomous political unit nor the seat of any important government. Decision over licenses for municipal services and governmental investments ultimately rested neither with Delhi's municipal committee nor with its British commissioner, but with the lieutenant governor of the Punjab government in Lahore.

Accordingly, when the Delhi committee formally resolved to invite Wilson's syndicate to apply for a license for its venture, their decision was immediately forwarded up the official chain for further consideration. First, the deputy commissioner sent the resolution to the commissioner, noting that the prospect of electric lights and a tramway for Delhi's citizens was desirable, but "there is no chance of the Municipality being able to embark on any such scheme *suo motu* [on its own behalf] for many years to come."[51] In his own note on the resolution, forwarding it on to Lahore, the commissioner, H. C. Fanshawe, was less enthusiastic, and he appended his reservations, both moral and regulatory, about the desirability of any such plan: "I have not given the subject of the introduction of electrical lighting in the East any special consideration, and I am not very sure that the same considerations apply here as in the West, or even that the conditions of Calcutta, Colombo, &c., with their large European population and special shipping trade requirements, would apply to a case like Delhi, in which all business comes to an end and all shops are closed shortly after dark. I mention this because I doubt if the Municipal Committee should pay for electrical lighting in Delhi any more than it already pays annually for street lighting."[52]

The specifically imperial ideology, with its peculiar combination of paternalism and utilitarianism, appears clearly in Fanshawe's terse reflections on the introduction of electric lights in the East. Indeed, the municipality's expressed desire to consider the syndicate's offer was here contradicted at the outset on the basis of an official's separate assessment of local needs, under the Orientalist understanding that Delhi had a natural diurnal rhythm unmodified, and unmodifiable, by technology. But Fanshawe also took care to specify laissez-faire principles of economic regulation as a further basis for his reservations, stipulating that he was "very doubtful whether we can or should give the concessionaries a monopoly of supplying electric light to private persons." Indeed, as the syndicate's application for a license to begin works traveled in higher circles of government, the core issues of concern to colonial officials were matters of economic prudence and the propriety of governmental versus private action in this field.

Questions about the propriety of governmental action—expressed over the terms and scope of any license to be granted for municipal electrification—long delayed provincial action. The legal precedent for such a concession was to be found in the provisions for earlier, horse-drawn tramways (never previously installed in Delhi), which had been installed and operated in Bombay and Calcutta by private companies under municipal licenses with long-term public purchase clauses. Yet, after several subsequent inquiries from the municipality over the course of a year about the prospect of a license for Delhi, J.F. Connolly, the revenue secretary in the government of Punjab, wrote in July 1902 to the commissioner of Delhi (after Fanshawe's retirement, T. Gordon Walker), explaining that the delay in government action on the Delhi license was due to the continuing negotiations being pursued (in the parallel context of the Lahore license) over just how the novel demands of electrical enterprises could be met under these precedents. He offered pointers for Delhi's municipal government in its own negotiations, specifically telling it to avoid any absolute monopoly and ensure that important business and residential areas were marked out for compulsory service. He enclosed legal opinions from the central Public Works Department on the terms of any license, including draft language from the central-government electricity bill that was then under consideration by the viceroy's Legislative Council.[53] Connolly, like several other of the officials involved in this bureaucratic wrangle, embraced the public promise of electrification to be won by thus encouraging private enterprise, writing that "the proposed scheme promises to be one of such undoubted public benefit and utility that its undertaking ought certainly to be encouraged by acceptance of all reasonable concessions asked for by the applicants."[54]

These civil servants, as might be expected, worry more over technical details, terms of compulsory purchase, and definition of obligatory services than they do over the social impact of electrification. At most, they aim to protect a hypothetical future public interest. Almost to a man, the officers of the central government were clear that no outright monopoly ought to be granted to the ultimate license holder, but that a "virtual monopoly" might be assured to the "first entrant" in the field, and that it was desirable—from the point of view of enticing further investment and moving forward with installations—to provide whatever such assurances were possible to businessmen willing and eager to invest in an Indian municipality.[55] The public works secretary of the central government was less certain, writing in some detail to the provincial officials to object to the conditions being requested by the syndicate, and noting that "capitalists will of course stand out" for the most favorable terms "as long as there is a chance of getting [them]."[56] But the force of the argument was on the side of granting the syndicate some sort of license on quite generous terms.

The syndicate, now represented by a Lahore barrister, Arthur Grey, wanted more than just a limited license, however generous its terms. Grey insisted on a legal monopoly or, failing that, guarantees that the syndicate would have a free hand to sell electricity and provide motive power within Delhi (and Lahore) for as long as possible, and he took his concerns directly to Sir Charles Rivaz, the lieutenant governor of the Punjab. Grey argued in a "demi-official" letter that only a clear governmental grant of monopoly rights would allow him to secure the necessary subscriptions to the issue of capital stock in London, on which the whole enterprise depended.[57] Lieutenant Governor Rivaz forwarded this note to his officials, agreeing that he "would be very sorry to see a scheme of such undoubted utility wrecked" by bureaucratic maundering.[58]

And there the matter of the installation of electricity in Delhi stood for some time, the government hamstrung by its own lack of legislative action but also, in a larger frame, caught at an impasse between the interests of the "public," as represented within the government in the form of concern about monopoly power and future terms of municipal purchase, and the demands of capital. For an entire year following Wilson's initial approach to the municipal committee in 1901, no formal governmental action was taken and no works were begun in Delhi. Further, even Sir Charles Rivaz's vigorous promotion of the syndicate's interests could not reach past the continuing official doubts—both about the terms of private action in this field, as expressed by the provincial and central governments, and about the propriety of electrification in the East, as expressed by Delhi's commissioners. In mid-1902, Delhi's Commissioner Walker noted that, given the uncertainties that had been raised by various officials, the municipality should wait to take action until the central government published a final electricity act (which would not happen until early 1904). His lack of urgency is striking: "It is of no real importance to the Committee, though it no doubt is to the would-be concessionaries, whether the scheme is undertaken now or a year later."[59]

With the public announcement in September 1902 that Delhi would be the site of a Coronation Durbar, however, it became clearer that electrification was of immediate importance to the municipal committee—the stature and reputation of the city, as a host to thousands of visitors and a showpiece of Britain's government in India, were at stake. On the outskirts of the city, the giant encampments were soon under construction, the light railway was laid to connect the main rail lines with the durbar grounds, and two Calcutta firms—Kilburn and Company, and Osler and Company—were installing electric power in the encampments and grounds several miles distant from the city itself. The municipality's awareness and anxious monitoring of the durbar works is conveyed in a letter dated October 15, 1902, from Deputy Commissioner Douglas to Commissioner Walker, letting him know that the municipality was hoping to have Wilson's company install some electric streetlights for the durbar. "The municipality are understandably anxious to get their street lighting in order before the forthcoming Durbar," Douglas wrote, and Thomas Wilson was ready to begin work immediately and had "agreed to install the lights by the 15th December 1902." This deadline fell just two weeks before the opening of the durbar, and authorization was needed to begin work right away, even if outside the terms of the still-delayed concession for more permanent works.[60]

These works evidently went forward, though the "giant shadow" of the durbar meant that there was a cessation of official correspondence about the Delhi electric license at this point. By April 1903, after the durbar had come and gone, the "Delhi Electrical Power Syndicate" was said to be "electrifying the city" under a provisional license, and the municipality had leased land to the syndicate for a first electrical plant.[61] This plant, installed to power some forty-five arc lights on the main thoroughfares and a modest number of private lights and fans, was the first true municipal electrification of Delhi.[62] The inadequacy of this initial installation to meet the needs of the city was almost immediately evident. In November 1904, a new deputy commissioner, Parsons, writes to Walker in dismay at the state of the lighting and ventilation in the city's municipal and judicial offices. "From Kashmir Gate to the old intra-mural European cemetery," he says, "there are government offices, a church, two schools, rest houses, post and telegraph offices, and a police station which would benefit from electric supply, and indeed such supply is available on Queen's Road, just across the railway tracks." Parsons notes that electric fans "are much more serviceable than hand-pulled punkahs and not very much more costly," and offers a picture of daily life in unelectrified Delhi much more various and full of activities than Fanshawe or Walker could apparently conceive: "In the telegraph and post-offices they work in the winter months very much by lamplight; . . . there are church services by lamp light[;] and also, in the kacheri [municipal offices], although they may not be used much . . . there are record office rooms and a hazir's *mal khana* so absolutely dark that lights are needed in the day time."[63] Parsons urges, on grounds of utility, the extension of the electrical works thus far at least. Yet there is no evidence of much more progress for nearly a year thereafter.

Not until 1905 did the syndicate finally receive what it had first requested from the municipality in 1901—a license to install and work an electric tram and electric lights in the city of Delhi. As it happened, the terms of the ultimate license, issued only after much more back-and-forth between Arthur Grey and the upper-level officials in the government of Punjab and the passage of the 1903 Indian Electricity Act by the viceroy's council, were not much different from those first requested. True, under the terms of the 1903 act, no positive monopoly could be secured, but the syndicate was still granted rights to install and supply electricity within Delhi for forty-two years.

More importantly, the "virtual monopoly" that had been mentioned in early negotiations over the terms of a license was explicitly invoked, one last time, in the background of this licensure process. Shortly before the licenses for Delhi and Lahore were finalized, and to ensure that they would be acceptable to the syndicate, a secretary of the Punjab government wrote to Grey with word from Sir Charles Rivaz, saying that he was "willing to give you an assurance that, so long as your Company fulfills the requirements of the public in a reasonable and efficient manner, he will not during his term of office grant a license to any other applicant in either [Lahore or Delhi], and will further leave on record his opinion that subject to the above condition your Company should not be hampered by interference in this manner."[64] The note of the communication is apologetic, but the effect is clear: the syndicate's virtual monopoly was given demi-official force in the form of a personal assurance.

With both license and this personal assurance in hand, Grey himself went to London and formed two companies, inviting British investment in the Delhi and Lahore projects. The Delhi Electric Tramway & Lighting Co., Ltd., was registered in London on May 16, 1906, and an announcement and invitation for investment was published in the *Times* on May 19. The company would, it was advertised, develop trams and domestic and industrial installations, and would take over "as a going concern the lighting installation which now supplies current to the municipality for street lighting, and also to the . . . Railway Station, and also to a few private customers." The company promised investors the secure prospect of return, because, "in a climate like that of India" the electric fans "must be kept in operation continuously day and night during the hot season."

This prospect of the thorough electrification of Delhi, day and night, was long unrealized, for a number of reasons, principal among them the financial weakness of the Delhi Company itself—a weakness that can be traced to the overlapping and murky financial interests of the various actors involved. The address of the new company in London was that of a solicitor who represented a number of other colonial electrification ventures—in Cairo, Baghdad, and elsewhere—and immediately upon capitalization the company contracted with the Edinburgh firm of Bruce Peebles & Company to manufacture and install the tramway and electrical plant in Delhi—a contract totaling some £103,000. In turn, the Bombay firm of John Fleming—our friend Thomas Wilson, principal and promoter—was hired by Peebles as its subcontractor in Delhi and as the operator of the existing concern while works were under way. When Bruce Peebles "suffered severe financial difficulties" and had to declare bankruptcy and abandon key contracts in February 1908, as a result of a failed attempt to build a light railway in Wales, the Delhi contract suddenly came into the hands of liquidation lawyers.[65] The local staff of the company in Delhi had to scramble to complete the works without much help or further capital from the London promoters. Meanwhile the Bruce Peebles liquidator, seeking to realize the value of the equipment that the company had already shipped to India, began suspiciously enquiring into the "relation" between the Delhi Company, Thomas Wilson (its principal in Delhi), and its Bombay contractors John Fleming & Co (which Wilson also managed).

As generators and other equipment destined for Delhi's expanding tramway and electric lighting system were tied up in the litigation over Bruce Peebles's assets, the director of the Delhi Company's retail operations wrote to the commissioner of Delhi in March 1908 to apologize for recent cuts in the electrical service to his home. "We realise that we cannot shirk the blame for this sort of thing, but hope you will bear in mind that we have suddenly been placed in an awkward position, by having to take over the complete running of the whole construction works and Power Station . . . with only the staff that were engaged by our late contractors."[66]

If these outcomes of risk and failure, accompanied by bureaucratic delays and false starts, "creative" corporate structures and instability of capital, are all characteristic of a novel technological enterprise, the colonial administration and its administration of culture still play a decisive role in the larger arc of electrical installations in Delhi. Municipal services in the colony were generally underdeveloped, according to the historian and onetime colonial bureaucrat Hugh Tinker, because of the hypertrophy of the apex colonial bureaucracy and the centralization of even mundane decisions. As he wrote in 1954, "Any account of [Indian] local government in the 1890's and early 1900's must deal with anti-climax and inertia."[67] As a more recent commentator has noted, the local colonial government in Delhi was systematically "underfunded and overstretched."[68] But bureaucratic bottlenecks and "red-tapism" do not exhaust the ways in which government routines hindered and delayed early electrification in Delhi. Bureaucratic delay would be unremarkable without the power granted to the very few officials at the apexes of power—the commissioner, the lieutenant governor, the viceroy, and ultimately, as we will see in the next chapter, the king-emperor himself—to supplement law with personal assurance and local decision, to enforce their doubt or register their enthusiasms, and to palliate the generality and force of the law with both private words and prolix pronouncements. In Curzon's hope for a sovereign boon to realize the moral grandeur of imperialism, as much as in the "assurances" of the lieutenant governor of the Punjab to a syndicate of British businessmen, we can discern the routinized charisma integral to the actual functioning of the bureaucratic structures of colonial rule. In these structurally significant actions we see the performance of certain kinds of power relations, both expressing the rationality of colonial government and its ritual form. This kind of bureaucratic charisma certainly affected the pace and temporality of electrification of Delhi. Again, this is not just a "secular" failure of governmental routine—electrification in Delhi, in its technical and its bureaucratic aspects, was also a productive site for the further ritual elaboration of ideologies of rule.

After the Durbar

Ultimately, efficient electrification was long deferred in the streets of Delhi, just as Curzon's hope for a permanent moral uplift from the ritual efficacy of the durbar was also unrealized. Still, durable effects can be credited to Curzon's durbar, especially in the emphasis throughout the durbar

Figure 4. Site of the viceroy's camp eleven months after the durbar.
Copyright © The British Library Board (India Office Records Photo 430/79 135).

on ceremonial proclamation as a mode of government—Curzon's denied boon reveals the rationale behind and the ideological legitimacy of the assurances offered to the electricity syndicate in Delhi. Ultimately, the durable effects of the imperial ritual are, paradoxically, most tangible in the very temporary character of the infrastructure installed for it.

Materially, little remained of Curzon's grand spectacle after the close of the durbar—its most visible legacy was the *absence* of the technological devices that once proliferated on the durbar grounds. This ritually produced lack of infrastructure appeared in the half-obliterated traces of roads and rails running across the land that had lately been occupied by the Central Camps, and in the open trenches stripped of their copper wires.

In the weeks after the close of the durbar, the tents were disassembled and wires were dug up, furniture and fixtures were returned to the suppliers or sold at auction. The electric plant that had been installed to power the central encampments, with all their illuminated tents, as well as to provide energy for the telegraph station that communicated the event to the world and the streetlamps that adorned the byways of the durbar grounds, was returned to the military. It was destined for a future, more practical use in a "great experiment" (in Curzon's own words): this power plant had been purchased in order to ventilate and light the barracks of the British soldiery of the Indian Army by electricity, so as to keep the "common British soldier" away from the "native bazaar" with its "low temptations."[69] This, like the auction of the rest of the accoutrements of the durbar—many of the elaborate dressings of the various tents and pavilions being bought by India's princes, in a further recycling of the trappings of aristocracy—all went to reinforce Curzon's commitment that this ritual should be conducted on efficient and businesslike lines.[70]

The promises of a ritual that would be both useful and efficient, that had governed the conception of the Coronation Durbar from the start, were in fact realized when the matériel that had enabled the temporary technological occupation of Delhi for these ceremonial purposes was sent off to new productive uses in military barracks, or simply sold at auction. Of course, the durbar improvements were always meant to be temporary, serving only a brief moment of glory, and resale of materials was intended to help balance the books of the show. But still, one might well be struck by the thoroughness with which the imperial state came and went, and wonder at what this temporary ritual "exhalation rising suddenly from the plains of Delhi" means for our understanding of imperial power, with its self-conceptions of permanence and strength and its tangible, still-resilient effects.[71]

As we have seen, at the heart of the durbar ritual itself was a glaring absence that was to become quite important. The lack of a sovereign boon left Curzon deeply disappointed in the outcome of the durbar. After the conclusion of the festivities, he wrote to his wife to complain about the sterile formalism of a rite without a boon. He said that he, personally, had gained nothing from the event other than an uptick in his established reputation as a great organizer, and regretted that he was seen by most as nothing more than "a magnificent state Barnum, an imperial Buffalo Bill," the maestro of a grand and empty ceremonial.[72] However, when a mere nine years later Edward VII had died and George V was planning his own coronation festivities, the new king felt it was necessary to meet and exceed the precedent set by the previous Coronation Durbar, and he determined to go to India to appear before his subjects there and offer in person some great boon to mark the occasion. The problem of what that boon would be was central to the preparations for the next durbar and its ultimate significance; this problem took on such importance, no doubt, because Curzon had so successfully promoted the idea that granting a boon was what imperial sovereigns did when they acceded to their high estate, and that this was an act of "Indian statesmanship." As we will see in the next chapter, the road from Curzon's empty pageantry to a historically and materially consequential ritual performance of British kingship was traveled in less than a decade—and the historical return of Curzon's ritual logics implies that they were never as empty as all that, after all.

The brief occupation and technological transformation of the plains north of Delhi laid down, temporarily, a structural pattern and set an ideological template for both technological regulation and ritualized forms of government procedure, complete with personal power and habits of delay and disregard, that would continue to structure techno-politics in the city—and would inspire nationalist politicians to formulate real alternatives to colonial beneficence with its temporary technological gifts. Curzon's durbar had these effects not despite but, in important ways, because of its lack of permanent technological traces or material reminders left behind on the terrain of Delhi. When the next durbar was being planned, nothing was left of the one held merely nine years earlier. The whole durbar city had to be rebuilt, on a yet more massive scale. Yet the earlier durbar set the precedents and instituted the ritual expectations that would shape all that transpired in that later event. Indeed, the immaterial traces left behind by Curzon's durbar can only underscore the power and importance of its ritual logics, as the very motive force that drove the machinery of imperial government.

RITUAL CENTER AND DIVIDED CITY

> The sacrificial battleground has been turned into a serene and perfectly
> ordered ritual emplacement. There, freed from the ties and contingencies of
> the world, the yajamāna [patron of the ritual sacrifice] strikes out in his own
> universe through the artifice of the ritual. But this artificial universe, for all
> its systematic control of even the smallest detail, is brittle and ephemeral. The
> price it must pay for the perfection of its order is divorce from the world.
>
> —J. C. Heesterman, *The Inner Conflict of Tradition*

The writer Nirad C. Chaudhuri records in his autobiography a "comical encounter" he had in New Delhi in the mid-1940s. It was wartime, the city was bursting at the seams, and the governmental center of New Delhi was engulfed by freshly built colonies of civil service accommodations. Seeking an acquaintance in one of these concrete lanes, Chaudhuri came upon a countryman, a fellow Bengali, and asked him for directions: "Passing along a line of buildings which looked like stables, I asked an elderly Bengali whether the clerk whose house I wanted to find lived in that row. He angrily pointed to the letter 'D' carved on the top of the building and said: 'Do you not see that these are D-class quarters, and the person you have come to see lives in E-class quarters?'"[1] Chaudhuri adds that "even my very superior clothes . . . did not protect me from the D-class disdain I brought on myself by being on visiting terms with a clerk who was living in E-class quarters."

The clerk whom Chaudhuri encounters speaks in the language of colonially ranked and graded difference and expresses as a lived sense of status the division and separation that organize his housing, his work life, and ultimately his sense of significant space; this is a mundane and modern expression of ritual distinction. Chaudhuri's own response is more ambiguous. When Chaudhuri registers his disapproval of the status-mongering of the "elderly Bengali," is he expressing, by contrast, a patrician disdain for this petty assertion, or a more democratic ethos of equality? Obviously, these last two options are not equivalent political affects (we should note that Chaudhuri points out his own "very superior clothes"). Yet there is a local, materially salient warrant for taking Chaudhuri's very record of this encounter as a claim on equality. The housing might be D-class (and therefore higher than E-class, showing the narcissism of small differences), but the street is still a public one, part of a larger urban network of communication to which the citizen has rights of access, and this is part of what motivates Chaudhuri's response to this encounter (remarking on it in print).

This kind of contradictory and conflictual encounter between principles of equality and distinction reveals a distinctively late-colonial and emergent-nationalist counterpoint in the political life of New Delhi. The city, its rulers, and its citizens, from its founding as an imperial capital, set performances of ritual unity against practices of division and distinction, and assertions of status against claims of equality. Bureaucratic or occupational distinctions such as those between the salary grades in the colonial services routinely do take on lived reality. That fact is an expression of the social process of class formation and the human one of status differentiation. But there is something distinctively "New Delhi" about this encounter as Chaudhuri relates it. A kind of haughty distinction-mongering was built into the conception of the city as an imperial capital, and even into the design and regulation of its technological infrastructures. Yet New Delhi's colonial founding, growth, and eventual urban integration of its region, infrastructurally and in terms of everyday life, also fostered, in the waning years of the British Empire, the emergence of a different understanding within its imperially organized spaces: an understanding of the commonness of urban amenities, and of a shared subjection to the imperative demands of the legal and technical devices of modern life. In this chapter I explore the ritual creation of New Delhi as an imperial capital in 1911 and particularly examine the growth of the new capital city's elaborate, overengineered infrastructures and the contrast they presented to the continued underinvestment in the provision of modern amenities in the newly "old" city next door. Overall, I explore the creation of political and social distinctions with modern technology and in urban spaces throughout the late-colonial period, and the countervailing, anticolonial thought that took these techno-political distinctions as objects of critique.

The King-Emperor and the Mahatma

The newly enthroned king-emperor, George V, and his consort Queen Mary alighted at Apollo Bunder in Bombay on December 2, 1911, on their way to Delhi for the second great durbar in nine years. Once again events of death and succession involving a distant royal family spurred massive interventions in the fabric of Delhi, and another durbar city was built on the old, vacated durbar grounds; a stately pleasure city, a miracle of rare device (in Coleridge's anachronistic, but apt, phrase), was decreed and duly appeared. But this time around, the showy temporary installations for a royal coronation would herald a permanent change in Delhi's urban fabric and a decisive shift in its historical fortunes. To mark the occasion of his coronation George V was planning to announce the transfer of the imperial capital from Calcutta to Delhi—effecting personally just the kind of royal boon that Curzon had been barred from proclaiming at his durbar nine years earlier.

It is true that such acts of kingship, of colonial beneficence, do not deserve the outsized attention sometimes granted them as key events in the political life of India, as such. Still, this ceremony did mark the beginning of a new phase in the urban history of Delhi. George V's very presence in India in 1911—he was the first (and last) reigning British monarch to visit India as emperor—indirectly spurred the creation of much of the urban fabric in which later, consequential events in Indian political life would take shape. Even the spot in Bombay where the king and queen arrived in India later became the site of an imperial monument, the Gateway of India, still a landmark of that city's urbanism but one that is now overwritten with myriad other (more demotic and democratic) myths and meanings.

The king-emperor's magnificent gesture at his durbar, transferring the capital of his empire, was kept secret right up to the day of its announcement by the king himself at the height of the 1911 durbar. It was decisive, for the high officials of the colonial state, that it appear as if the king-emperor himself was shifting the capital and creating a new tie of beneficence between the Crown and the empire. While Curzon had to argue strenuously for some boon at his durbar in the face of official constitutional scruples, those who planned the 1911 durbar met with no such reservations—rather, this time imperial officials seem to have embraced the idea; they saw a great durbar as an occasion to efficiently effect major and consequential changes in the terms of colonial government (though this time around the ritual form would be more a screen for, than an intended vehicle of, policy changes). In short, the king-emperor's act of kingship was based in the very ritual logics that had been earlier defined and defended as necessities of "Indian statesmanship" by Curzon. These logics then took technological, architectonic form in the imperial capital city that grew south of Delhi in the years after 1912, rehearsing colonial distinctions between British and Indian, ruler and ruled.

At the same time, however, such imperial ideologies and their material instantiations alike were challenged by late-colonial Indian politics—most importantly by Gandhi's separately developed moral critique of technology and of the toll it imposes on any shared, convivial life. In *Hind Swaraj*, his major nationalist tract written in 1909, Gandhi addressed his fellow Indians passionately, telling them that they must understand that "machinery is bad" and give up the apparent utilities of railways, tramways, and electric light (the latter was, he wrote justly, not even seen in English villages).[2]

These are the poles of the argument pursued in this chapter: at the one end, imperial ritualism and its centralizations, distinctions, and discriminations, and at the other, Gandhi's creative counter-ritual practice and his oppositional vision of an Indian constitutional order based not in great centralizations and charismatic actions but rather in minor ethical practices repeated, endlessly, throughout the whole political body.

It is important to underscore at the outset that the thoughts and works I describe in this chapter are plural and, moreover, responded to rapidly changing technological and legal conditions over the course of thirty-odd years. Gandhi was acute in his diagnosis of the times in which he lived, and he was neither naïve nor entirely utopian in envisaging an alternative to industrial and technological urbanity still immanent within existing social formations in 1909. By the 1930s, however, political and technological realities on both sides of the colonial relation were different from what they had been at the end of the first decade of the twentieth century. In no small part because of Gandhian mobilization, constitutional reforms had opened up pathways of partial participation in government by nationalists; meanwhile, a goal of cheap, plentiful, ever-present electrification was now unquestioned by colonial officials and electrification appeared to be the most urgent project for any country seeking to take its place on a world stage. No more doubts remained about "electrification in the East," even though India's municipal, industrial, and electrical institutions alike remained underfunded and overextended. Colonial officials and nationalists in government, despite real budgetary limitations, had abandoned laissez-faire principles of ideal competition and privately funded improvements under the canopy of imperial beneficence, and instead promoted unitary municipal ownership of electrical utilities as the most efficient means toward the new goal of comprehensive electrification of urban life.

Such shifting judgments are collective, historical products, and I treat them as such here—I make no claim of historical priority or causal relation. For example, imperial ritual itself did not spur Gandhi to thought or action, although the despotic and autocratic habits of imperial administrators certainly did (apart from some fleeting comments in his autobiography, Gandhi paid scant attention to the durbars of 1903 or 1911). But the contrast of ritual logics, between the durbar and Gandhi's own political-ritual practice, is interpretively fruitful for understanding the political meanings and moral lessons that different actors derived from the material and social conditions of urban modernity in Delhi and New Delhi throughout the whole arc of late-colonial electrification. Ultimately, I argue, the political thought expressed by the durbar and by Gandhi's critique of technology both converge on core modern problems, continually posed within technological modernity, of the ritual means and the social ends of government.

Durbar Day, Again

After their arrival in India and a ceremonious progress from Bombay to Delhi, the king and queen settled into an opulent tent and awaited the advent of the Imperial Durbar. On durbar day, they arrived at the new amphitheater "in the Imperial state in which their loyal subjects love to see them," escorted by "a great cavalcade" of horsemen and "clad in their superb coronation robes and [wearing] imperial diadems," to borrow the breathless description of the *Times of India* correspondent.[3] Escorted by the viceroy, Lord Hardinge, the royals arrayed themselves on a dais to receive homage and honor from their Indian subjects. If Curzon's durbar had been a spectacular show, this one more than trebled the scale of that one—three-quarters of a million people were estimated to be in Delhi specially for the 1911 durbar, with some three hundred thousand housed in the official encampments alone.[4] Like nine years earlier, most of the costs of the durbar, including the crafting of the bejeweled imperial crown that King George wore, were borne by the Indian treasury—on the theory that most of the results, in terms of political stability and increased loyalty, would benefit India herself (the crown was later kept in England, however).

Indian elites and soldiers massed around the focal point of the crown and its bearer in the amphitheater, and ceremonial proclamations were read out as the king looked on. As in 1903, this durbar was the occasion for a series of special remissions and small gestures of grace proclaimed in the king's name by the viceroy—including grants for extension of public education and surplus pay for members of the Indian Army. However, once the viceroy had read out his list of official and relatively minor boons, the heralds trumpeted again and the king-emperor himself stood forth and began speaking. As specially printed gazetteers were distributed to the crowd, the king proclaimed: "We are pleased to announce to our people that on advice of our Ministers tendered after consultation with our Governor-General in Council [the viceroy], we have decided upon the transfer of the seat of the Government of India from Calcutta to the ancient Capital of Delhi, and simultaneously, and as a consequence of that transfer, the creation at as early a date as possible of a Governorship for the Presidency of Bengal."[5] The brief royal proclamation continued, specifying other administrative changes to be effected "as a consequence" of the transfer of the capital to Delhi, and then the king took his seat again.

What did this strange interruption in the expected ceremonial routine mean? The question was asked around the amphitheater as word filtered out of what the king had said. The capital of India was to be moved; India's historic center was to regain its rightful preeminence over mere administrative or commercial centers like Lahore or Bombay. Meanwhile, in a curious conjuncture, the king had also announced that his government would undo the partition of Bengal, an unpopular move made six years earlier by Lord Curzon. The backstory to this proclamation helps flesh out its immediate impact, and its significance for the subsequent planning and construction of New Delhi.

In 1905, Curzon had undertaken an administrative reorganization that split the presidency of Bengal, with its nascent Indian-run municipal and legislative institutions, into two parts along religious lines. This was widely recognized as a gambit to diminish the power base of a newly assertive (Hindu) Bengali nationalism concentrated in Calcutta, as well as an effort to manipulate communal divisions between Hindu and Muslim Bengalis in the imperial interest.[6] By 1911, the partition had become a political grievance of wide resonance—nationalist agitation had increased, and the Bengali-led *swadeshi* movement had emerged as a vital nation- and empire-wide political force that drew much of its energy from this injury to Indian self-determination. The presence of open political agitation, focused on the issue of partition and self-determination, however, made it difficult for the British (because of their own ideas of imperial protocol) to find a pathway to reversing the partition, however much that might be a desirable policy goal. "British colonial policy," as the historian David Johnson has pointed out, "could never be seen as stemming from or influenced by nationalist demands." In order to reverse the partition of Bengal, "a new colonial policy was needed, one in which the reunification of Bengal was seen as a secondary component."[7] A grand act of kingship suited perfectly this need for flexibility within the straitjacket of colonial prestige—in the colonial context, it was now politically strategic to allow the king to perform certain actions in the ritual context as if they were expressions of his own grace and concern for his subjects.

But this was not only a matter of cynical manipulation, of colonial raison d'état exploiting a fortuitous coup de théâtre: the idea of a boon was integral to the very conception of a grand durbar as it had been laid out by Curzon. As soon as the notion of a durbar to celebrate George V's coronation was mooted, historian Ronald Frykenberg has noted, "the very first serious question raised pertained to the finding of a suitable boon."[8] As Frykenberg has shown, reversing the partition of Bengal was early raised as a possible boon in internal debates over the king's planned visit to India, and yet the publicity problems this presented and the imperatives of royal ceremonial led a minor official on the viceroy's staff to propose the rather more spectacular idea of transferring the capital entirely away from Calcutta. Combined, the two actions were uniquely suited to the demands of the day: one had the supposed benefit of appearing to be a gift to Muslims—the ancient capital of the Mughals would retain its former glory—and this, with its kingly splendor, would outshine the imperially dubious concession to mobilized Hindu elites in Bengal.

In the midst of all this official calculation over the royal boon, no serious question was raised about its British constitutional propriety—in part because great secrecy about it was maintained to protect the impression that the king's autonomous action was all that motivated these decisions. While cabinet ministers were involved in the planning of the boon, the leaders of the opposition in Westminster were not informed of it until the night before the durbar.[9] The attractive idea of a royal boon, of a great act of "Indian statesmanship," seems to have justified circumventing normal constitutional procedures. Ultimately, the king's personal pronouncement was more than just a frill of imperial ceremony; it was essential to the intended *political* effect of the Imperial Durbar that the Crown appear in its personal guise, as if law were being made by the king himself.[10] For perhaps the last time in British constitutional history, a reigning monarch pronounced as an act of kingship a series of consequential legal changes, without any prior public legislative consideration (while retaining the verbal forms of ministerial responsibility). Most importantly, once the king personally had announced these changes, they were effectively fixed in stone. As the *Times of India* correspondent wrote, in admonitory response to official arguments over the wisdom of these new policies, "This is the King's will, and it is final."[11]

Comparative Darkness

The logic of this particular boon drew on and reinforced a set of imperial assumptions about the monarchical potency of the British Crown—able to communicate across the imagined cultural divides that separated colonizer from colonized, and to tie far-flung political communities into a single structure.[12] The Crown was certainly a much-invoked symbol of unity in imperialist writing; more importantly, it was also central to formal, constitutional thought on the relations between the several parts of the British Empire. The Crown provided a cynosure around which (racially defined) debates on self-determination in the dominions and the character and scope of colonial dependency revolved, while the Crown's status within the domestic government in Britain also spurred debates about legitimate forms of political power and constitutional order in a modern, industrial, class society.[13] Royal protocol and the staging of kingship buttressed a certain imperialist imagination of solidarity at the very same time that they ratified the categories of colonial difference and reinforced a class interest in the empire.

With an act of kingship, then, this polyvalent British Crown was installed in Delhi, and it was decreed that a new city should be built to accommodate it. This gesture of monarchical grace and unity also raised to a new significance the pattern of uneven relations between imperial rituals and urban processes, between colonizing pomp and colonized city, that had been such a feature of these rituals in Delhi. The contrast between durbar city and historical center as it had been staged in both 1903 and 1911 was to become a permanent opposition—multiply mediated in technological and in civilizational terms—between an "old" and a "new" city.

As we will see, in future years the imperial capital's lavish material provision for stately sumptuousness and technological display came to contrast sharply with the old city's "backward" and "dilapidated" infrastructures. This pattern of temporal and technological contrasts was set by the very organization of the durbar itself. The Delhi municipal committee noted this in early 1912, in a draft report on the just-concluded durbar. "The lighting of the main roads was undertaken by the [central government's] Durbar Committee," they wrote, "and most of the municipal electric lights became superfluous and were extinguished." They went on, in a passage that was crossed out in red pencil on the draft preserved in the official archive, and which is suppressed in the final, printed version of the report, "Had the temporary arrangements been combined with the permanent municipal electric light system . . . the municipality would have been able to make, on favorable terms, a permanent extension of its electric light system on many roads that for want of such an arrangement relapsed into comparative darkness as soon as the Durbar was over."[14] Apparently, the passage that pointed out this missed opportunity was judged, by some censor, as striking the wrong note—perhaps it was even thought to be somewhat disloyal—for it was not included with the otherwise fulsomely congratulatory reports in the final record of the durbar. Yet the municipal

committee's description of a comparative darkness offset by the brilliant
ceremonial lights of the durbar was to be prophetic when it came time to
install more durable electrical infrastructures for the new capital. Let us
now turn to the details of that story.

Doubled Urbanism

The city of New Delhi did not simply appear, in a single day. It took years
of planning by the architects Edwin Lutyens and George Herbert Baker
to map out and realize the grand imperial city that had barely even been
imagined by the officials who devised the transfer of the capital as a purely
political gesture. Yet within a year of the king's pronouncement, the government moved to a temporary secretariat in the existing European quarter or "civil lines", and soon overflowed the available office space and
housing in Delhi—putting great strain on the infrastructures of the city.
With the sudden arrival of the whole imperial state, with its mundane
and its ritual requirements, electric supply soon became a major concern
of local officials in Delhi. In fact, instead of any extension of the service
already provided throughout the historic city and into the civil lines by
the Delhi Electric Tramway and Lighting Company, separate electricity
works for the government offices were shortly begun.[15]

To clear the legal way for these imperial works, the company's license was
revised and narrowed, to include providing electricity service in the existing
city districts only. With the laying out of the new capital city on a site south
of both the old durbar grounds and the historic center of Delhi, a modern,
municipal electricity plant was located beyond one end of the major ceremonial axis that would become Kingsway (now Rajpath). At the same time, the
local government also began to exercise more oversight of the operations of
the Delhi Company, its trams, and its supply of power to a customer base in
the old city. In 1914, the Delhi Company came under increased scrutiny for
its slapdash work methods and its failures to "ensure the public safety," and
an electrical inspector and an inspector of trams were appointed as regular
government officers.[16] Curiously—perhaps out of an excess of regard for the
terms of the original license, by then just a decade into its tenure of forty-two
years—no effort was made to unify, either technically or organizationally,
the electrical operations in the old and new Delhis.

In this regard, Delhi's technological history during the building of the new capital contrasts sharply with that of the major coastal cities of Bombay and Calcutta, or even such colonial entrepôts as Singapore. In each of those cities, after an initial period of experimentation and competition for control over electrification and municipal transport, single institutions soon took over power supply and tramways in the entire city and its region (though these were not publically operated concerns in Bombay or Calcutta). A unity of technical and of political control (whether through a municipality or a single monopolistic utility) was already, by this time, seen as the most "advanced" mode of utility supply. In Singapore, according to the historian Tilman Frasch, the drive toward unification of formerly private ventures under one urban authority meant that the city's "commissioners tolerated the Singapore Electric Tramways Company for as long as they were unable to run their own system to generate electricity, but got rid of the whole business at the earliest opportunity."[17] While in other colonial cities the story of electricity is one of increasing efficiency, integration, and technical coordination—with a prevailing logic of regional integration and urban growth—in colonial Delhi and New Delhi the primary political and technical reality was reiterated, reinforced *division* between different spheres, official and nonofficial, despite the fact that a slow and piecemeal infrastructural integration of the two cities was under way, under pressure of urban necessities, from the start. That is to say, a technical and a political gulf separated the customers of the Delhi Company, living in the areas still remaining within the jurisdiction of the Delhi Municipal Committee, from those who lived within the boundaries of the New Delhi Municipal Council, with its centrally funded infrastructures. The electrical infrastructures in the "old" and "new" cities, throughout the imperial period, remained very distinct in their disparate form and organization across the divides—mostly legal and bureaucratic, but also aesthetic and technical—that separated the ceremonial city from the commercial one. New Delhi's public supply was installed as a high-standard 220-volt alternating current system; the old city—which still shared its electricity supply with the tramway—continued to operate on its direct current system for decades.

The "stately" requirements of the new capital meant not only that the electrical supply there was more technically modern, but also that all its electric installations were lavishly overengineered to better fit with the imperial grandeur of the surrounding buildings. "The electric system of

New Delhi was not designed solely with regards to business principles," according to the report of a 1937 government commission on electricity supply in both the old and new parts of the city. "A glance at the roadside lamps alone would indicate this. From aesthetic reasons they are of a type in which a considerable part of the illumination is absorbed by the glass of the lanterns. Similar reasons . . . compelled the adoption throughout New Delhi of the underground system of distribution mains."[18] This resulted in immediately apparent visual differences between the different parts of the city—tramways with overhead wires continued to ply Old Delhi's streets, and electricity poles strung with wires marched throughout the suburbs outside the official capital, while around the stately buildings of the ceremonial core one saw only imperial monuments.

This kind of simultaneously aesthetic and technical disparity—as well as the experienced variation in levels of service across the two cities—provided the kinds of contrasts that nationalist intellectuals took up and transformed in their political thought: a clean, smooth, and orderly electrical modernity set against a dilapidated "old" city, "backward" because of its lack of power. Jawaharlal Nehru's much later, rueful reflection on India's technical level of development (discussed earlier) is anachronistic but not inapposite to cite once more here. Speaking to the Constituent Assembly, he noted that the course of India's history was partly determined by its "failure" to develop its energy resources, "and we became backward because of that." He goes on to highlight the "enormous new power" that was being exploited in other countries around the world.[19] In Delhi, such comparisons did not involve great imaginative leaps—the contrasts between the two cities of old and new Delhi, in levels of energy development and technical standards, were tangible and close at hand.

The modern infrastructures of the imperial city were under the command of the central state—providing, also, a massive demonstration of the developmental, technological powers that state officials, were they willing, could summon forth. Old Delhi and New Delhi, in their close proximity, encouraged a double vision of urban problems, prospects, and solutions, inculcating a sense that only political independence could ameliorate such material disparities and propagate a higher standard of technological functioning—and ultimately freedom—for Indians.

In the late 1920s a group of residents of the old city, including Delhi's nationalist member of the local municipal council M. Asaf Ali, formed the

Delhi Electricity Consumer's Association. The group petitioned the municipal government to support improvements of supply, and intervened in legal disputes between the Delhi Company and its customers over rates, installation costs, and quality of service. The association took a decidedly political stand on the question of whose interests ought to be served by such an urban utility—they complained to the local government about the company "usurping the valuable rights of the public," especially by contrast with the efficient and low-cost electricity supply available in the official parts of the city.[20] Nearly a decade later, the official commission on Delhi's electric supply would, in turn, note that the high rates of the private company were a frequent topic of "protests in the [Delhi municipal] legislature and elsewhere."[21]

After long experience dealing with Delhi's bifurcated electrical infrastructures, ultimately, even colonial officials took to heart the lesson that technological provision of common, modern amenities required unified control and the exercise of political power. While the residents of the old city addressed the local government with complaints against the private company that was charging them high rates and engaging in "high-handed" business practices, the imperial state prided itself on what it had achieved in New Delhi, and how. The New Delhi Municipal Council—an entirely official body—in its representation to the power supply commission serenely prided itself on its own modernity and efficiency in embracing the supply of municipal utilities as a government function. "Municipal trading in general utilities is becoming an increasing feature of all modern communities," the council wrote, "and [the municipality of] New Delhi may wish to continue what we have so well begun in this direction and deprecate any attempt on the part of private capitalists to divert to their pockets the profits accruing from the running of a public utility concern."[22] At the time the council wrote this, the Delhi electricity supply business beyond the confines of Lutyens's ceremonial city was still in the hands of "private capitalists," though there were fewer and fewer profits to divert.

The New Delhi council's claim about the bright future of municipal control over electricity utilities reveals a shift in techno-political logics within the local state in Delhi, which was also reflected within the imperial government as a whole. By the time New Delhi was finally inaugurated in 1931, it was already an example of the efficacy and even moral purpose of a new kind of state power that provided basic, infrastructural necessities, and dynamically adjusted relations between the public good and private rights. This new capital was something more than just a showpiece of imperial power; it also promoted and promised, by its very existence, a vigorous role for state power in the future material and moral transformation of the colonized society.[23]

Ritual Divisions

The complicated questions of monarchical beneficence that both the durbars featured underscore the importance of these ritual events, not only as occasions for the display of the ranked-and-graded colonial sociology on which imperial ideologies depended but also as effective moments for the exercise of real power. This is difficult to describe without deploying the language of antidemocratic "distortion" and unconstitutional personal power, marking any deviation from formal proceduralism as illegitimate. But while there is much to be noted here about colonial despotism, in these contexts it was strictly speaking constitutional proprieties that were at issue: when it came to the personal embodiment of the Crown in the colonial context, all of the ritual of monarchical boons and sovereign grace was perfectly proper and even *constitutionally* necessary—or could be legitimately argued to be constitutional. Curzon might have been a particularly pedantic student of ritual, but his reputed ability to take himself seriously in the midst of great Oriental pageantry was, in fact, one of his main qualifications for imperial government. As the historian Michael Mann has said, Curzon was "the most outstanding imperialist ever," largely because of his consistent production of statements and performances that communicated this theory of government.[24]

Still, the grand gestures of the colonial hierarchs crafted brittle, friable distinctions and differences that presented real barriers to political action and urban community. Guided by ritual protocols and the constitutional proprieties of imperial rule, the social life of the new city fostered, in addition to material divisions in the design and layout of urban infrastructures, separations of status and belonging—as we saw with Nirad Chaudhuri's encounter with D-class distinction and disdain. In the terms of a Durkheimian description of social solidarity, the technical divisions created in the imperial city by ritual pronouncements and concern for propriety marked out domains of quite literally mechanical solidarity—a turning inward toward similitude and internal elaboration of structure that barred any wider connection across these differences.[25] The imperial capital of New Delhi continued, as it grew, to embody these kinds of mechanical divisions in its very institutions and infrastructures, and in its hieratic separation from the processes of the wider region. Its high technological standards went hand in hand with ritual elaborations of rank and hierarchy, just as its "geometric design reflected the hierarchical nature of its official society."[26]

This hierarchical nature was often elaborated with great pomp and display. Every February, the social season reached its apex with "Delhi Week," during which the same kinds of entertainments that had enlivened the durbars were staged again, such as a horse show and polo tournaments, and notables from all over the empire partied in the purlieus of the grand imperial city. "At Prince's Place, the sound of saxophones wailing the tune of 'I'm Forever Blowing Bubbles' wafted from open loggia windows beneath a floodlit dome," in the somewhat overripe description of the historian Robert Irving. "Pennons in state colors snapped above the hoods of the sleek limousines, and imperious honking from deep-throated klaxons announced the arrival of rajas at a fellow ruler's festive tamasha."[27] The ritual city that was reanimated, periodically, by these events, amid the infrastructural installation of luxurious devices and erection of sandstone palaces, was lauded as a masterpiece of modernity as much as a center of ritual order—while never losing its appearance as a kind of shimmering, temporary "exhalation" of principles of mastery and order.[28] The architectural historian Sten Nilsson observed, for instance, that the views of New Delhi preserved in photos from 1922 show that it had the "character of a permanent camp" and hence represented a built realization of the ritual logics of the durbars.[29]

Although the imperial city increasingly became one with the old city, its roads, railways, buses, sanitation, and water being used by all of Delhi's denizens, the divisions and distinctions that separated old from new *electrically* remained in force. Even as ideological currents (and technical rationality) tended to favor municipal control and operation of public utilities, the colonial government's durable concern with protocol, propriety, and distinction seems to have forestalled more radical reforms in Delhi's infrastructures. In 1939, when the various municipal and central government

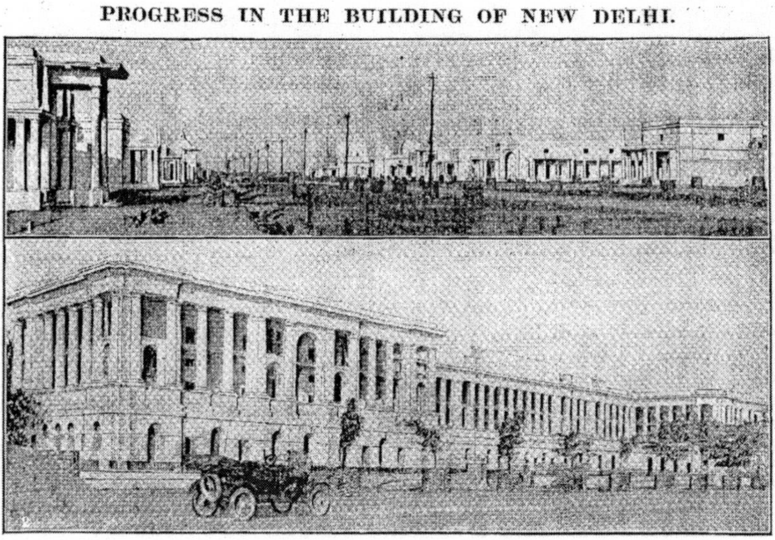

PROGRESS IN THE BUILDING OF NEW DELHI.

Figure 5. "Progress in the Building of New Delhi," *The Times*, February 23, 1922.

electricity institutions operating around Delhi were finally combined under one unified public board, the license of the private electricity concern was still not extinguished—it was merely amended, once again, to limit the responsibilities of the company to reselling centrally produced power to its existing customers.[30] On these limited terms, the Delhi Company—now known as the Delhi Electric Traction and Power Company—survived until independence.

That decisive political watershed arrived at the very moment that the company's initial license was coming to its "natural" end—that is, the license had been issued in 1905 and secured a forty-two-year concession to provide electricity and operate trams, so it was due to expire in 1947. This was, of course, a moment when much larger issues were on the table, and the future of Delhi's municipal institutions could no longer be a narrow, technical question among bureaucrats but rather was now really a constitutional one, to be addressed in the nation's new Constituent Assembly. We will pick up that thread in chapter 3. But first, let us turn to the development of Gandhi's ritual and ethical challenge to colonialism, which also involved creative ritual reworking of the technological terms on which the imperial state governed everyday life.

Ethical Technique

No one discerned the fateful connection between technological mastery, ritual distinction, and imperial rule more clearly than M. K. Gandhi. Before New Delhi was even announced as an idea, Gandhi was writing about ritual-cum-political distinctions and impositions as the key to colonial rule and identifying colonialism's vaunted technological-cum-civilizational mission as a devastating arrogation of a wider social power to shape and manage relations. In his richly allusive, dialogical writing in *Hind Swaraj*, composed in 1909, Gandhi directly and critically addressed the costs imposed by the speed of modern transport, the bodily ease provided by modern technologies, the thoughtless massing together of bodies for work and politics.[31] "Good travels at a snail's pace," he wrote, and "it can, therefore, have little to do with railways. . . . It may be a debatable matter whether railways spread famines, but it is beyond dispute that they propagate evil."[32]

Gandhi built his critique from many then-ongoing conversations across the English-speaking world and beyond about the haste and wastefulness of modern civilization. Many intellectuals, not all of them pastoral romantics, lamented the encroaching presence of technological urbanism and industrialism on the edges of even the most rigorously local, village- and community-based consciousness. A quarter of a century later Ruth Benedict would note that a whole critique of "our civilization" was based on the characterization of modern societies and their inhabitants as "thoroughly extrovert, running about in endless mundane activity, inventing, governing, and, as Edward Carpenter says, 'endlessly catching its trains.'"[33] Gandhi's view of technology built explicitly on this strain of thought, mentioning Carpenter's works by name.[34]

In a perceptive reading of *Hind Swaraj*, Ajay Skaria has argued that what Gandhi ultimately challenges in that tract is the unlimited ambition not only of machine civilization but of all imperializing aspirations toward mastery of others and of the world—technologically or otherwise: "For Gandhi . . . the self is always already inclined towards infinitude and mastery, towards the inevitable and indispensable machinicity which reaches its apogee in 'modern civilization.' . . . Central to his politics, thus, is the struggle to limit that infinite and machinic self which participates both in domination of the other and its own subordination"[35]

Materially speaking, the devices that Gandhi embraced as elements of an ethical and political remaking of Indian life—most famously, the spinning wheel and the handloom—demanded discipline, limitation, concentration, inner-worldly centering, and also entailed a social decentralization of energy and effort. By contrast to the techniques deployed in the symbolic instrumentation of a "durbar city," the humble machines of Gandhi's ritual practice were all but impossible to use in a project of accumulation, aggrandizement, and ritual splendor—and this was their power. The *charkha*, or spinning wheel, in particular, was simple, exigent in that it demanded concentration and skill and utterly tied production to a place and time. Decentralization rather than centralization, distribution of effort and yet unity of purpose were the political entailments of the technical practices that became the symbols of Gandhi's movement—and they stood in frank opposition to the proliferation of insignia, demands of pompous display, and central ordering, division, and separation of ranks and grades that were integral to imperial ritualism.

Gandhi subsequently made this *counter-ritual* practice—spinning— central to the public identity of the Congress Party. This was a masterstroke of symbolic politics. There was nothing "traditional" (on the terms that the colonial state laid down for understanding that idea) in the work of spinning. It was work, it deployed (humble) technologies, and it focused the mind while producing something all-but-elemental in its simplicity (thread), rather than rehearsing the colonial stereotype of mindless tradition and its endless reproduction of elaborate forms of ornament. Spinning evaded all reference to the "obligatory frame of custom" by which the colonial subject was encouraged to recognize himself as subordinate within the governmental hierarchy of the imperial state (I am thinking of the kind of performances of pomp and regality enjoined upon princes, to which Gandhi morally objected).[36]

Gandhi's practice opposed, negated, and reconstructed ritual distinctions of the sort on which the imperial state thrived and by which it governed. Indeed, he explicitly spoke of producing "prestige" from modest means, and this was a key goal of the "constructive program" that he wrote out in 1941, in anticipation of independence. "Let not the reader make the mistake of laughing" at any simple practices such as spinning and wearing *khadi*, which were laid out in the constructive program, "as being part of the movement for independence," Gandhi wrote. Participation in the independence movement was itself, he clarified, a kind of office, an insignia of honor, and this is what makes great the little task of spinning:

> Many people do many things, big and small, without connecting them with non-violence or independence. The same man appearing as a civilian may be of no consequence, but appearing in his capacity as a General, he is a big personage, holding the lives of millions at his mercy. Similarly, the charkha [the spinning wheel] in the hands of the poor widow brings a paltry pice to her, in the hands of a Jawaharlal [Nehru] it is an instrument of India's freedom. It is the office which gives the charkha its dignity. It is the office assigned to the constructive programme which gives it an irresistible prestige and power.[37]

Gandhi's political genius was to counter-invent—with spinning—a ritualism of bodily technique and social technology, one that forestalled any capture within the colonial state's own performances of technological, traditionalizing pomp, and yet produced (in the form of homespun *khadi* cloth) a new "livery for Indian freedom."[38] His ritual innovations emerged from his direct reckoning with both the ritual and the technological terms of the imperial state. It is part of Gandhi's modernity, his correct assessment of the power of the colonial state, that he chose *archaic* but still *technological* symbols to instrument and convey his message about the relations between social power and freedom within a modern political order.[39]

Constitutional Powers

The colonial durbars influenced the form of the great imperial city, as cultural distinctions between colonizer and colonized were built into the capital of imperial rule as technological divisions. The different technological and material standards apparent across the divide between Old Delhi and New Delhi were described—in government reports and in mundane bureaucratic communications—as inevitabilities of the demands of "state," that is of the protocol and custom that in the imperialist imaginary underwrote constitutional government. Gandhi's response to this ritually reiterated principle of division and separation, as Nehru once pointed out admiringly, was to apply "an ethical doctrine to large-scale public activity."[40] But this means that Gandhi's ritual insights were political, too—not only ethical, but acting, like their imperial cousins the ritual logics of the durbars, on a larger scale to shape distributions of power. Ultimately, Gandhi's ritual practice was not only inspired by an ethical insight into the failings of imperial rule; it was also founded on his insight into the constitutional force of the technologies of speed and communication that were used as the medium of colonial power and granted particular form to its hierarchies and separations. He saw how these technologies, as they were built into the very ordering of the colonial society, shaped the distribution of power and regulated the possibility of social action—and he explicitly objected to the centralized control and hierarchical exclusions from participation that electricity networks necessarily entailed (whether they were used to power floodlights on a dome or to drive a humble sewing machine).[41]

Gandhi's religious, ritual thought and action depended, no less than the imperial sociology, on an insight into the canalization of effort and the instrumentation of social relations in modern states. As the anthropologist Milton Singer observed, "Gandhi's campaign for weaving hand-spun *khadi*," although not a practical program for large-scale production, "does seem to have succeeded in dramatizing concern for cottage industries, the dignity of hand labor, and village underemployment, while providing a symbol for a self-respecting cultural identity in a successful mass political movement." Making explicit the conjuncture of ritual practice with modern technical and political imperatives, Singer concludes, "From this point of view, one might argue that the *charkha* (spinning wheel) really articulated in an archaic idiom the voice of Congress and of the sewing machine."[42]

Over time, the ascetic religious and ritual practice that Gandhi invented as a political repertoire, along with the zero degree of ceremonial accoutrements in his own personal adornment, had decisive mobilizing effects and helped shape the very conditions for the postcolonial Indian state.[43] The imperial durbars had been designed to convert technological splendor and imperial pomp into legitimacy and bonds of loyalty, and to realize one particular constitutional vision in the form of personal power aggrandized by great but isolated (and often temporary) installations of splendid ritual technique. Gandhi sought to challenge this constitutional order with a counter-ritual practice that opposed, term for term, the ritualism of the colonial state and thus could form the basis for an alternative distribution of power, opening a radically different set of possibilities for regulating social action. After independence, partition, and the death of Gandhi—when a real constitution had to be written for the newly independent state in the turmoil of active political contest—his vision was hard to install in its purity and ethical rigor; somewhere in the conversion between his ritual idiom and modern, institutional expression of power something was sure to be lost. But the inspiration provided by Gandhi's core question remained: Can ethical goals of freedom and political goals of wide-scale participation be not only realized, but also emblematized, in the technological devices of an urban, industrial modernity?

Though Gandhi's answer was negative, his counter-ritual and counter-technological practice was nonetheless based on a core and durable insight into technology and power: political communities are always in some way divided and separated, and to realize any form of unity one needs rituals and devices that can invent and reinvent social power athwart material distributions and divisions. At their best, such devices will produce possibilities for mutual relation, adjustment, and change.[44] The conceptual alternative to a symbolic and political technology, such as Gandhi's, which is both ethical and endlessly open to transformation, is one that imposes imperial isolation upon totalized groups (as colonial ritualism and its despotic separations did, ultimately shaping further distinctions between E-class and D-class Indians). This imperial logic of division was made plain in the final recourse to partition as a governmental technique at the end of the colonial era. Partition replicated on new and horrible terms the kinds of separations and divisions that had characterized New Delhi's growth alongside the old city. Once again territorially and technologically separated groups were opposed to each other, and the powers of each could only grow at the expense of the other's vitality.

Part II

NATIONAL GRIDS

Political integration has already taken place to some extent, but what
I am after is something much deeper than that—an emotional integration
of the Indian people so that we might be welded into one, and made
into one strong, national unity, maintaining at the same time all
our wonderful diversity.

—JAWAHARLAL NEHRU, speech in Bangalore, October 1955

3

The Lifeblood of the Nation

A railway engineer is active in solving problems involving motion and fuel,
but he also dreams of all that can also be set in motion thanks to this
solution: not only travelers, merchandise, minerals, and soldiers,
but also ways of living, ways of communication, modes of
distribution of the possible and impossible. . . .

It is, however, hard to describe at the same time the technical conception
of the new system of combustion and the dreams of those who conceive it,
who themselves make up the difference between what depends on them
and what they must leave up to others.

Isabelle Stengers, *Thinking with Whitehead*

The 1930s opened in New Delhi with the inauguration of the finally complete architecture of the imperial city. Another durbar-style ceremony was held on February 13, 1931, but it was a muted and minor one, for the times were not propitious for self-aggrandizing pomp or monarchical boons. Gandhi had launched his civil disobedience movement the year before and was soon to be imprisoned for his political activities. Indeed, one might think that the whole juncture bespeaks a massive failure of creative political imagination on the part of the imperial government.[1] At the same time, the future prime minister Jawaharlal Nehru was involved in a personal political education that involved intense, ongoing questioning of the place of India in modern world history. In the early 1930s, during his own periods of political imprisonment, Nehru wrote long letters to his daughter, Indira, comparing the current condition of India to historical and technological developments throughout the world, and situating the movement for Indian independence in a worldwide picture of political and social change. A cosmopolitan by temperament—unlike the equally well-traveled Gandhi—Nehru shared with his daughter (and with a wider Indian audience, through the collection and publication of the letters) detailed images of technological progress and the political conditions that fostered it, making politicized comparisons between India, the United States, and Soviet Russia throughout.

Summing up in one letter the nineteenth century and the changes it rang in world civilizations, Nehru recited a litany of change-producing forces at once intellectual (Darwin and Marx), political ("democracy and socialism"), and fundamentally human (famine, war, civilization). "The dominant feature of this era was," he wrote, "the growth of capitalistic industry by large-scale power production—that is, production with the help of some mechanical power, like water, steam, or electricity." He added, parenthetically, "we have the name 'power-house' for an electricity-generating plant," punning on the convergence between the political transformations wrought by industrial capital and the harnessing of industrial energies.[2] In later letters, Nehru drew a more explicit political lesson from recent global history: he contrasted India's still-feeble harnessing of industrial energies with the high "general standard" of technological progress reached in other, uncolonized countries and identified India's lack of national freedom as both cause and consequence of its backwardness. He particularly dwelled on, as a model for India, the ongoing attempts in Soviet Russia to transform social life through large-scale electrification. He wrote, about the postrevolutionary government of Russia, "Lenin started a huge scheme for the electrification of rural areas, and mighty electric plants were put up." The political intention of such a program was uppermost in Nehru's mind, and he ascribed a beneficent and generous spirit to Lenin's electrification program: "This was meant to help the peasants in many ways. . . . The peasants, whose villages were lighted up by electricity and much of whose farm work was done by electric power, began to get out of the old ruts and superstitions and think on new lines. . . . Lenin was so keen on his scheme for electrification that he used a formula that became famous. He said that 'electricity plus Soviets equals socialism.'"[3]

What Nehru knew, or did not, about Soviet plans and their implementation is less important here than the relation he envisages between transformative power and expanding political consciousness. Building on this connection, Nehru argued throughout his writings and in his practice as prime minister that large-scale social change, ameliorating conflicts between differing mentalities and ways of life and ultimately producing freedom and democracy, was best achieved by directed, planned harnessing of industrial power and its orderly distribution throughout the country. The Congress Party planning committee set up, with Nehru as chairman, in the late 1930s repeated this thought, when it described electrical energy as the "very life blood of the industrial nation" and therefore the most important tool of India's forthcoming independence: "Social ideas regarding generation and supply of electricity have undergone a complete revolution in all countries of the world. From a luxury article which it was in 1903, it has become a domestic and industrial necessity comparable with water supply, and is an essential element of modern life and civilization. Electrical energy is something more than a commodity; it is the very life blood of the industrial nation which must flow abundantly and without interruption if the nation's strength and well-being are to be preserved."[4]

For the experts and lawyers on this planning committee, as for Nehru, the technical instrumentation of a modern, industrial economy was a matter of politics, solidarity, and laws—of "social ideas"—as much as it was one of economics or technical design. Moreover, the relationship between political freedom and technology was, as these writings indicate, a *vital* one—not a matter of merely material changes or technical "scaling up" of capacities, but of transformative forces that, in the proper political conditions, could (and should) be extended across the whole of the social fabric to produce a wholesale change in the strength and well-being of the people. This vision was one of coordinated, comprehensive, and most of all expansively inclusive technological progress, political participation, and moral advancement. Within this discursive formation, electricity was always marked out as an index to the progress and prosperity of a country—and as the most urgent area for any government activity intended to build a more prosperous, more dynamic polity. The scientist and journalist Meghnad Saha regularly argued against colonialism and for public control and development of electric power in his journal *Science and Culture* (published from 1935 onward), writing, "If consumption of electricity *per capita* is taken as a measure of civilization, India stands almost at the bottom of the scale . . . [and] no government which neglects this line of activity can be said to discharge its functions properly."[5]

The political lesson was both clear and widely shared at the time. Saha specified that "almost every civilized country in the world has come to regard the supply of electricity as a public utility concern and has taken steps for full development of its power resources, for adequate control of the production of electrical energy and for ensuring the public of a cheap supply and protecting them from profiteering exploitation."[6] At the same time, as we have seen, colonial officials in New Delhi also argued for the unification and municipalization of all electricity supply in the city, on the basis that "electrical energy is a phenomenon which has perhaps less regard for parochial divisions, compartments, and attitudes than almost any other feature of our modern world."[7]

By the mid-1940s, with independence on the way, New Delhi became the seat of a national government occupying the terrain and architecture of the imperial city. The life of the city was deeply affected by the rapid populational, cultural, and political shifts accompanying the independence and partition of British India. Delhi as a whole had a rapidly growing population of refugees and political migrants of all stripes, requiring urgent administrative responses.[8] In view of these transformations in the status and the human reality of the city alike, plans were set in motion to manage the growth of the city, link it to its region, integrate and improve its infrastructures, and provide for massive new governmental institutions. As a capital city, a new, national New Delhi would serve as a model for nationwide political efforts to institute coordinated control over the economic and industrial processes of a modern polity, and a political laboratory for working out questions of cultural identity and the meaning of political autonomy.

This second case study, focusing on constitutionalism and social and cultural integration in independent India, thus moves us away from entirely official and stately rituals in order to explore more explicit and contentious political debates—including, in the next chapter, arguments among anthropologists working elsewhere in India in this period over the political meaning of electrification and technological change. Through such political and scholarly debates, we may discern yet again the routine combination and ritual interaction of political ideals and material objects, this time in the making of a national "modern social imaginary" (to borrow Charles Taylor's evocative term).[9] I examine, first, how and why political control over modern infrastructures and their technical operation together formed a single problematic social reality and object of thought in this era, and, second, how technical devices came to be the prime material object around which a new understanding was forged of belonging to the nation and participating in its progress. In this second section as a whole, I trace in more specific detail how technological capacities and devices provoked thought among political actors in Delhi and beyond about "how we continuing stand or have stood [in relation] to others and to power"—and how this led to one specific articulation of a "modern theory of moral order" in the context of legislative and constitutional debates, as well as decisive transformations in the symbols of belonging that were available for appropriation and new relational uses elsewhere in India.[10] This is not to say that such "modern" imaginaries did not fully exist before independence, or that they have not been inspired by diverse cultural sources and political currents over time. Rather, the problems posed by urgent crises besetting local infrastructures, and efforts to plan for citywide and nationwide expansion of technological capacities, together spurred reflection upon core issues of citizenship, participation, and belonging at this particular juncture in India's political life. The contrasts made, the terms used, and the institutions imagined—and some actually realized—in this period continue to shape debate over the distribution of power and the terms of belonging in urban India to this day.[11]

Throughout this section, politicized contrasts and comparisons are essential to the meaning granted to technical questions and problems. Not all these contrasts put India's own progress and achievements in an unfavorable light, or emphasized her "backwardness." M. Asaf Ali, whom we have earlier witnessed being dazzled by the show of lights and of public, urban life in Marseille during his youthful journey to Europe, was in London in 1946, on his way to the United States as India's first ambassador to that country. This time, however, the experience of postwar austerities, including shortages of fuel and indeed of light, presented a stark contrast with both his earlier visit to Europe at the height of Britain's imperial splendor and with his own hopes for India's brilliant future. Asaf Ali remarked in his diary and in letters to India on the lack of power in this otherwise victorious capital. He particularly noted "the peculiar experience of walking through the dark corridor of the India Office by candle light."[12] His sojourn in postwar London reinforced his earlier conviction that Indians would have to build a new model of modernity on their own cultural pattern.

The shadows now engulfing the center of empire could not help but make India's future seem all the rosier. It was as if the wages of imperial despotism, so long paid by India, were now falling upon the imperial power, while the task of protecting and fostering the future—not only the national future but also a new international order—was at last to be assumed by India's political men and women.[13] In a letter to Nehru, Asaf Ali said, "Sitting here in London and getting an idea of the various bottlenecks which are holding up the rehabilitation of the U.K. after the severe blow which they have suffered, it is plain to me as daylight (which has been totally absent throughout my stay here) that the economic argument [for political unity and centralization] is the most potent and effective one."[14]

This letter helps indicate that the arguments made by Indian nationalists on behalf of a strong centralized government, one that would wield control over "public" things and stand in command of the industrial processes of a modern economy, involved more than simple admiration for the material achievements of industrial states, or straightforward ideological views about the role of the state in capitalist modernity. For Indian modernists like Asaf Ali, Saha, or Nehru (and the whole suite of experts and intellectuals who surrounded the political leaders and formed a new class of officials), statistics about the electrical (and material) poverty of India and its notional "failure" to develop modern energy technology were "a cause for moral reproof" (as the political scientist Sunila Kale has put it) and demanded coordinated and thoughtful leadership.[15] By the same token, as the urbanist Ravi Sundaram shows in his account of Delhi's postcolonial development, urban infrastructures and networks of technical integration could be celebrated as "statement[s] of sovereignty." The institutions of planning and development of infrastructures promised, to the intellectuals of the new India, "the mix of technology, information, and secular magic [needed] to propel modernization."[16]

The Electricity (Supply) Act

Independence, as a real prospect, arrived hectically in India at the close of World War II. With the establishment of an interim government in 1946, a Constituent Assembly was also elected on the basis of the old, limited colonial legislative franchise (its representativeness was further limited when, with partition, the Muslim League withdrew from India's new institutions). This assembly was gathered both as the legislature for the provisional government and to prepare for independence by writing a new constitution for India's political future. This dual purpose meant that substantive questions of economic and social regulation were considered in the very same body—meeting in separate sessions—as deliberated over the formal division and allocation of powers within the new state. In both its constitutional and its legislative capacities, the Constituent Assembly worked to build the legal and administrative armature by which, in the words of a contemporary observer, "a dynamic conception of law as the science of social engineering" could be operated.[17]

Indeed, throughout the government, the time seemed propitious for assumption of direct and centralized control over the country's industrial and technical capacities—very much as Asaf Ali had written to Nehru. A team of colonial civil servants helping to plan for the transition to a new, independent government reported to Nehru, "Not only in India but in many overseas countries there is a growing demand for the nationalization or public ownership of public utilities."[18] The physicist Homi J. Bhabha was asked to hold consultations with representatives of the electrical industry and develop a national program for coordinated development, and B. R. Ambedkar, as the interim minister of mines and works, introduced an electricity bill into an early session of the Constituent Assembly. Ambedkar's bill would have effectively centralized control over electricity at the national level; but that bill, delayed by external crises and internal pressures, never saw a vote.

A revised Electricity (Supply) Bill was introduced in the legislative session of December 1947 and was passed in early 1948, one among a suite of legislative acts of the assembly pertaining to major sectors of scientific and technical enterprise. In introducing the bill, N. V. Gadgil, the new minister of public works, made once again the case for state control over public utility concerns, citing the growing consensus among experts that such control was a way to secure and foster the public interest. The bill, however—as was pointed out in the debate that ensued—did not include provisions for full nationalization of electrical enterprise (a pathway just then being pursued, as India's legislators were aware, in Britain). Nor did it envisage the creation of a single national grid under central control. The constitutional and the political situation both militated against such "advanced" legal and technical provisions.

Under the 1935 Government of India Act, then still operating as the constitution of India, electricity regulation was an area shared between central and provincial governments—it was a matter of "concurrent" legislative competence. In this, electricity was unlike railways or posts and telegraphs, for instance, over which the colonial government had retained exclusive lawmaking power. In the 1935 law, electricity was treated more like professional licensing, factory inspection, and other more purely municipal regulatory functions that were partly devolved to provincial assemblies but with oversight and general policy still emanating from the central government. This concurrent status was, in part, a simple recognition of the provincial disparities in electrical development and the wide array of organizations, public and private, operative in this productive sector of the economy. The Electricity (Supply) Bill maintained this legal situation, as did the eventual constitution.

In support of the bill's moderate provisions short of full nationalization, Gadgil said that "the path of wisdom," as he adopted it in this draft legislation, was "the line of least resistance." This included creating provincial institutions to "coordinate and stimulate electrical development" within their boundaries and to exercise financial and technical control over existing commercial undertakings in cities and towns. In this way, "the investor is reasonably assured of a fair return and the benefits arising out of limited profits are secured to the consumers in the form of the low cost of electricity." Still, even this relatively minimal legislative program for provincial integration of electricity enterprises had already raised, Gadgil said, a great deal of passion on the "issue of nationalization versus private enterprise."[19]

One member from Bombay endorsed the government's "path of least resistance" and criticized his more politically radical colleagues (B. R. Ambedkar among them) for pushing immediate nationalization or even just further centralization than was already contemplated—the goal, he said, was to improve electrical installations, not merely to own them. Making an analogy with regulation of telephones, he asked if it would be satisfactory "if all the telephones in the country belong[ed] to the State," even if they did not work. "Apart from sentiment, money, and everything else, the actual facilities and convenience which add so much to the comfort and happiness of people are, according to me, the real essence of nationalisation."[20] Gadgil picked upon this rhetorical opening to strengthen his argument in favor of a progressive, but incremental, trajectory of subject-forming, citizen-producing legislation, of which this bill represented just the beginning: "If we thoroughly and efficiently control the [private] concerns in every respect," he said, "I have no doubt that gradually an atmosphere will be created where people will work more in the spirit of social service than for the sake of mere money or profits. They will develop a better sense of citizenship than they possess today."[21]

In her account of these legislative debates, the political scientist Sunila Kale highlights the symbolic importance of electrification to nationalist elites, and remarks that still there seemed to be a lack of political will for more extensive control over and centralization of electrical enterprises. She contrasts the Electricity Act of 1948 with the arguments made by Saha, for example, for regional and even national integration of electricity in a "grid system," and with the earlier legislative program for nationalization and integration proposed by Ambedkar. These more comprehensive visions represented a modernist pathway to technical integration and social equality that was, Kale notes, more coherent and consistent than the eventual legislation's accommodation to existing constitutional arrangements and economic and social interests. Ambedkar's "vision for [electrical] development was calibrated to a national scale to ensure that vital resources were shared across the national space," in the name of equality, "rather than hoarded in particular locales."[22] Kale astutely observes that coordinated regional development across state lines was hindered by this constitutional relegation of electricity to the state or provincial level. However, the constitutional settlement represented by the Electricity (Supply) Act of 1948 also reflected a deep connection believed to exist between electrical development and local freedom, between public control of shared utilities and the incremental growth of moral powers of self-regulation and self-reliance— the development "of a better sense of citizenship."[23] Commenting on India's nascent constitutional order, the Bengali Marxist Dhirendranath Sen used, tellingly, the phrase "municipal freedom" to characterize the material aspects of self-government in a worldwide capitalist system.[24] Just such local freedom, and how it might be realized through political control over urban infrastructures, became problematic in very instructive and revealing ways in the course of further debates over the constitutional status of the city of Delhi itself.

Sovereign Delhi

The years of the Constituent Assembly's existence were difficult and changeful ones throughout India, but the strains of independence were especially evident in the immediate urban environment of Delhi—the city's population grew enormously in the decade between 1941 and 1951, and its whole social and material character changed as it became a metropolis of nearly two million people.[25] While the assembly met in the grand central hall of the colonial parliament building in New Delhi under a constellation of bright lamps specially installed for the chamber's new and serious purpose, social upheavals, urban crises, and breakdowns of both civility and infrastructure were daily realities in the wider city. The changing fortunes of Delhi were immediately evident to those in the debating chamber.[26] Moreover, the urban rioting and states of emergency that periodically paralyzed life in Delhi in 1947 and 1948, the leading historian of India's constitution-writing process records, "brought home the lesson that local law enforcement and local—even provincial—government could be frail reeds in times of great distress, that the centre must have the power to preserve order and the processes of government."[27]

However, the experience of Delhi also offered other, contrasting lessons. If the debate over the electricity bill revealed tensions in and objections to the modernizing, centralizing legislative and constitutional program of the dominant leaders in the Constituent Assembly, doubts about centralization and unified government authority were even more evident when a question arose about Delhi's governmental structure—in particular whether the central government should continue to exercise legislative and police power over the whole city.

How was Delhi, as a national capital and now a major and modern city in its own right, to be governed in independent India? This question was initially raised in the course of a routine debate over provisions for "minor" or outlying administrative divisions—whether, and for how long, they might be put directly under central legislative control as their status in the new federal structure was worked out. A new version of an existing clause in the draft constitution had been introduced to clarify that the central government would retain the power to legislate for "minor administrations" indefinitely, and that these administrations included not only small states but also Delhi itself. This apparently small and technical question, meant to deal efficiently with an array of foreseeable problems and the ongoing crisis in Delhi, raised "practically a storm in the house."[28]

Over two days, the debating chamber was strewn with countervailing proposals for the future government of Delhi. Ideas for Delhi's new political status included granting it independent and sovereign status as a city-state, making Delhi the capital of a new Punjabi state, or leaving New Delhi as a government enclave and making the rest of the city a part of the neighboring (Hindustani-speaking) United Provinces. Each of these proposals was meant to recognize a different political, cultural, or linguistic constituency with stakes in Delhi's future status: to ratify the city's historical associations with Muslim political identity and autonomous sovereignty, to make it a new urban capital of Punjabi culture after the loss of Lahore to Pakistan, or to integrate it into the Hindi-speaking belt. Delhi's representatives also had their own stake, apart from their cultural or linguistic identity, in maintaining their political constituency intact.

Opening the debate on the amendment at issue, Delhi's Deshbandhu Gupta (a Punjabi lawyer and Congress representative) noted that Delhi had already long suffered from its status as an imperial enclave, pointing out in particular that no provincial-level patronage apparatus encompassed Delhi, and its young men therefore had no easy pathway to a government career. Given its size and centrality, he asked, should Delhi not be granted the status it deserved? "Delhi has already got a population of about twenty lakhs [i.e., two million people] and this is bound to go up further in a few years' time. Thus it makes a very good unit to be treated independently."[29]

Gupta further argued that the conditions of other cities, like Washington or Canberra, that were founded as governmental enclaves, could not apply to the case of Delhi and New Delhi, taken together. Although it was true that New Delhi had always been an imperial enclave, the historic city of Delhi had also been transformed and enlarged by the arrival of the imperial capital, and no fundamental split could now be introduced between them. "The population of both the cities is intermingled. Transport, electricity, water supply and all other essential services are common to both and even the population is common," Gupta said. Indeed, this portion of Gupta's speech is worth quoting at length in order to observe how he derives his moral and political argument for Delhi's unity, and for the legitimacy of its democratic aspiration, from the coherence of the city's material, infrastructural fabric:

> Sir, I have heard some people say that Delhi is much too small a place and that the demand for autonomy is being made merely to satisfy the aspirations of some local political leaders. This is a very cheap jibe, if I may say so and cannot be taken seriously. Such an argument could be equally applicable to our demand for self-government or independence in a wider sphere. I can assure the House that it is not as a matter of luxury that the people of Delhi have demanded autonomy or a measure of self-government or a voice in their administration. Their difficulties are real. . . . Several Ad Hoc bodies have been appointed like the Improvement Trust, the Joint Water and Sewage Board, the Delhi Central Electric Power Authority which have got official majorities and no effective representation of the people of Delhi. They plan and take big decisions about Delhi, but the people of Delhi have no effective voice in the administration of these bodies. . . . Some people believe that Delhi has benefited from the location of India's Capital here. Let us examine this. It is the right of every big municipality to own, control and fund the essential utility services like electricity, transport, water-works, etc. and they form a big source of their income. Do you know that they have never been entrusted to the Municipality of Delhi? The fact is that Old Delhi has been made to serve as a maid to New Delhi, which has been built as a Capital.[30]

The debate continued to the next day, with representatives making heated claims and counterclaims to represent the best interests of Delhi, from the basis of their own particularistic belonging to it. A member from the United Provinces argued that Delhi ought to be subordinated to his province's government, precisely because it had for so long been a "subservient" municipality, its political life governed by colonial habits of deference.[31] But the debate over Delhi also drew in representatives from farther afield. Ram Narayan Singh, from Bihar, stipulated that the issue of Delhi's government was of general concern because of a material experience of commonality that ought to touch the hearts of all the members of the assembly. He said, "There cannot be any question that we at least drink the water and breathe the air of Delhi . . . [and] it is the duty of the Members to see that . . . justice is done to her."[32]

This debate is significant for both the emotive tenor of the argument— the appeals made to collective freedom, bureaucratic ambition, and self-determination—and the specific and contrasting proposals mooted for the city's future status. Taken whole, moreover, this debate restaged as a municipal matter—explicitly as a question of control over resources and infrastructures—much the same question of "belonging" and self-determination that had already structured the independence movement. In his speech, Gupta himself raised the remarkable notion that if Delhi's administration had been reformed earlier, perhaps by amalgamating it with nearby towns and villages, the political weight and assertiveness of a Muslim-majority Delhi province would have been greater, and "perhaps the country would have been spared the agony of the Partition of India." That speculative notion aside, the problem of reconciling central legislative control over Delhi with the democratic demands and communal claims of the city's diverse populations certainly presented a test case and a kind of model for the wider working of the constitution, materially and practically, across India's diversities. Deshbandhu Gupta's claims on behalf of Delhi's residents, asserting their need as city residents (above other identities or modes of affiliation) for access to bureaucratic careers and for control over their own municipal institutions, indicate the high political meaning that infrastructures and public services had gained as tools and sites of collective self-determination. Moreover, as N. V. Gadgil had pointed out, introducing the Electricity Act in earlier legislative sessions of the Constituent Assembly, the hope and the promise of such local modes of participation was that practical control over the machineries of the urban economy would expand civic consciousness. Ram Narayan Singh echoed this idea in the course of the debate over Delhi's constitutional status. He argued, "Delhi being the Capital of India people from all parts of the country continue coming into or going out of Delhi. Therefore we should establish here an administrative setup that may produce a salutary effect on the whole of the country."[33]

At last, the arguments offered on behalf of "home rule" for Delhi particularly reveal the modernist, materialist, techno-political imagination that animated India's constitutionalism as a whole—both its reach and its real, material limits. Autonomy and self-government were associated with the improvement of a citizenry's material well-being on a broad scale; but belonging and participation were also figured at a local level as access to and engagement with the institutions of a technologically defined urban commons and public sphere. Moreover, the political-theoretical alternatives of centralization and decentralization, coordinated control versus home rule, are not statically opposed in these debates. The meaning of each shifts when seen from a different level of political ambition and participation. From the perspective of the national movement, constitutional government from a national center and access to the power of a governmental, bureaucratic administration were a means of freedom and self-rule, while from the perspective of the dependent municipality these very administrative means became evidence of material domination and continued, quasi-colonial expropriation of local democratic rights.

Gupta's strong plea for an autonomous, integrated, and democratic government for Delhi was not, at the time, to be effective—Delhi was included in the constitution as a "Part C" state, and ultimately (in 1956) became a "Union Territory" still subject to direct legislative control by the central government (after a brief period of home rule between 1952 and 1956). In fact, Delhi only much later—in 1991—won a legislative assembly almost on par with those of other Indian states. For much of its postcolonial history the dominating institutions in Delhi's material and political life continued to be arms of the central government, including some newly founded ones.[34] A prime example of the latter is, of course, the Delhi Development Authority, which was set up in 1957, administered by the Union Government's urban development minister, and often served the central state's projects of urban glorification more than the interests of Delhi's residents.[35]

Still, despite Delhi's special status as capital, the constitutional debate over its autonomy and its integration as a political unit gave expression to all-India political realities that were played out in the subsequent (and in some cases ongoing) struggles within India's federal system over state reorganization, and were also evident in the relative separation maintained by state electricity boards in later decades. That is, an abiding tension between, on the one hand, local realizations of autonomy through participation in infrastructural institutions, and on the other large-scale coordination of productive energies toward national freedom, was woven into the very constitution of the Indian state. This tension is what was interpretively at stake in the minor debate over Delhi's status. As has been often argued, the splendor and symbolic force of the "temples of the New India" (Nehru's phrase for dams, power plants, and other massive infrastructures) helped to legitimate and entrench the statist economic policies and centralizing politics of Nehru's specifically modernist vision for India.[36] But if, in the immediate period of constitution-building, this tension would be resolved in one way, symbolically and practically, alternatives remained available for the future renewal of local imaginations of self-determination.

Nehru's Symbols

In the years of constitution making, and while grappling administratively with the immediate impact of partition, Jawaharlal Nehru himself was involved in several alternative plans and projects for Delhi and indeed for the urban future of North India. In fact, in the Constituent Assembly he personally responded to Gupta's original amendment to provide for greater representation and autonomy for Delhi. After the debate, Gupta withdrew his amendment and Nehru assured the chamber that in due time more democratic arrangements for Delhi's government would be established, and that he and the Congress Party intended to do justice to the city's ambitions.

Throughout this period, Nehru was in regular correspondence with the American regional planner Albert Mayer, who was working in the United Provinces on rural development projects. In the spring of 1947 (prior to partition), Mayer wrote a detailed letter to Nehru advocating expanded use of regional planning in the government of independent India and stressing the refreshing difference from colonial routines that this would represent. This was in the very early days of the refugee crisis and communal violence that would mark independence, and Nehru wrote back to endorse the idea but also to acknowledge the impossibility of making any large new demonstration of public power in the present circumstances. "As you must know," Nehru wrote to Mayer, "we are in the midst of a political crisis and possibly on the eve of considerable changes in India. It becomes a little difficult to plan about anything unless the picture becomes somewhat clearer. Nevertheless, we cannot just wait for events and must prepare for the future . . . [by] planning for a happier and more prosperous India."[37]

Planning was, on these very terms—promising happiness and prosperity—among the most attractive tools of government for the Congress leader. Later Nehru and Mayer both would be key players in bringing American sociologists and urbanists to Delhi to draw up a master plan for the city.[38] The advent of master planning under international guidance would ultimately be one of the many ways that power over Delhi was concentrated in the hands of unaccountable, central officials. Yet in the early correspondence between Nehru and Mayer, the idea of planning (and more particularly, of an urban plan for Delhi) was as often mooted in terms of its communicative and symbolic importance—to stage and display the democratic legitimacy, progressive purposes, and material beneficence of the new government of India—as for its rational utility. The political message that planning would send was a central part of the wider meaning and purpose of the exercise. In 1950, Mayer explicitly proposed an expansion of a Master Plan already underway for Delhi as an "important example for future India, and indeed an important world example or pilot undertaking."[39] At this time, moreover, the problem of the "emotional integration" of nations was a topic under analysis broadly within international institutions like UNESCO, and it was soon to be taken up by the Indian government's education department explicitly as a matter of political communication.[40]

In 1953, after a brief visit to New Delhi, Mayer wrote again to Nehru with a novel idea for the "symbolic crystallization" of the "new independent New Delhi . . . in place of the old colonial New Delhi." Mayer proposed a "planning-and-architectural" solution to what he and others saw as the mismatch between India's independent institutions and the imperial grandeur of its inherited capital city. He envisaged a thorough reorganization of the ceremonial core of New Delhi to convey the meaning and importance of the new constitutional order and also to help glorify the state's technical apparatuses and institutions. "There are new democratic foci of power," Mayer wrote. "It seems to me that this must now be expressed. We have the Parliament, above all. There is the National Planning Commission. When I got this far in my thinking, I felt we must have one other element, approximately equal in status, to accomplish the new independent symbolism that seems so necessary. Tarlok Singh told me that there was also going to be a new Supreme Court. That news really electrified my thinking, because that trilogy *is* the essence of the new national India."[41]

The inclusion of the planning commission as a "focus" of democratic power may surprise but is consonant with how planning was talked about at the time—and comes of course from an American planner and New Dealer who believed strongly in the political benefits of coordinated control over regional resources. Nehru wrote back within a week to say simply that "I like your general approach to this question but, frankly, we cannot spare large sums of money for magnificent buildings, as much as I would like to have them. I was not aware that it was proposed to put up a building for the Planning Commission." Ultimately, these institutions were never coordinated into one composition, one image of state power, although the Planning Commission did erect a large new building to the side of one of the axes of power along Rajpath, and the Supreme Court was granted a new home to the north of the ceremonial core.[42]

Mayer may not have much cared whether or not the institutions to be entrusted with India's democratic future were directly representative or not. But his firm advocacy of "planning-and-architectural" solutions to problems of representation and participation within the new constitutional order echoes many other voices of this era, in which questions of democracy were so often articulated in material, infrastructural terms. A moving ratio between material interdependence and political autonomy is at stake in all these efforts to grasp the meaning and the promise of technological modernity, and to give it local institutional or material expression—from Nehru and Saha's admiration of massive national electrification, to Deshbandhu Gupta's definition of Delhi's governmental and civic aspirations in infrastructural terms, to international collaboration in planning.[43] In their effort to produce new sites, new material arenas, for political action—with the goal of national or emotional "integration" foremost in mind—the disparate actors in the constitutional and technological debates of postindependence India also helped figure new moral bonds, and hence also the possibility of acting across material differences within the nation, shifting that moving ratio between interdependence and autonomy.

The Poetics of Technology and Statecraft

In an influential reading of the postcolonial politics of planning and development, Partha Chatterjee has focused on Nehru's political speeches and writings to argue that the high politics of independence-era India were utterly disconnected from the true spiritual and social needs and social imagination of most of its citizens.[44] For Chatterjee, the postcolonial modernist investment in planning, bureaucratic control, and the infrastructures and goods of a state-led capitalist economy served in large part to concentrate power in a Congress-led government. Chatterjee argues that ultimately the freedoms of democratic citizenship and the social promises of inclusion that came with independence were both sacrificed to a purely material definition of prosperity and participation alike, a capital-friendly vision that effected a passive revolution over the moral community of the nation, with its distinct aspirations. Vernacular political life in the new nation was subordinated to the power of a unified party and a centralized state that deployed "modern statecraft, and the application of technology" to ameliorate poverty but in the process expropriated rather than empowered the mass of citizens.[45] Chatterjee's argument—taken as a historical thesis about Indian state formation—is powerful, and empowers critique; but it is important to note that he makes a strong contrast between *meaningful* Gandhian mass mobilization and *instrumental* Nehruvian planning and centralization in order to underscore his claims about the postindependence, developmentalist, social-welfare state. For Chatterjee, Gandhian mobilization depended upon a "religious semiotic" that touched the very depths of Indian consciousness and fostered hopes for freedom; by contrast, Nehru—distant from the masses—offered material hopes of prosperity to mobilize partisan support and legitimate his government.[46]

This contrast as Chatterjee makes it, and his judgment about Nehru's political practice, resonate with Hannah Arendt's famous attack on the intellectuals of the French Revolution, whom she charges with sacrificing revolutionary dreams of true political freedom in order to address the material needs and poverty of the masses—in the process, aggrandizing the revolutionary state, freeing it of any of the tethers that might have bound it to preexisting moral norms and moral community.[47] Motivated by a distanced pity for the state of the masses and concerned only with their poverty and dispossession, Arendt says, the French radicals sought to redress material woes, to address the "social question" through mobilization of public power, but in so doing produced an unanswerable justification for whatever acts they saw as necessary to defend the state since they had made it the only source of collective life. Arendt's point has been picked up and extended to the Indian case by Uday Singh Mehta. Mehta highlights the Constituent Assembly members' "deep and anxious preoccupation with social issues such as poverty, illiteracy, and economic development."[48] Directly quoting Arendt, Mehta says that the political lesson of this preoccupation is clear: "Any attempt to solve the social question by political means leads to terror."[49] On these terms, Mehta argues, Indian constitutionalism "must and does constitute power" rather than freedom, and only centralized power and autocracy were—or could ever have been—produced by high political actors' efforts to stitch the nation together by "giving expression to the existential needs of the unity whole."[50]

Mehta and Chatterjee both charge that the central, powerful state produced by postcolonial constitutionalism was buttressed by the very moralization of material need that was such a prominent theme of political debate and granted the constitutional debates their "arduous collective intensity" (Mehta's phrase). Judging the state's efficacy and attainments by the standard of development and material progress, the Constituent Assembly lost sight of the political life, in its varieties, of the Indian people. To borrow another phrase from Arendt, "It was . . . the ocean of misery and the ocean-like sentiments it aroused that combined to drown the foundations of freedom."[51] We might recall in this regard that the constitutional historian Granville Austin felt that the prime lesson the drafters of the Indian constitution took from the volatile political environment of Delhi was that local, democratic institutions could be "frail reeds" in times of great distress.

This critique of Indian constitutionalism has much to recommend it, as a guide to the postindependence politics of planning and development and the centralized structures it helped foster. Returning to the passages by Arendt that Mehta reads in one way, however, reveals that another interpretation is possible.

Arendt herself cites the very same phrase of Lenin's about communism meaning "electricity plus Soviets" that Nehru turned to as a guide to the technical and social promise of modern state power. She renders it (more faithfully, I think) "electrification plus *soviets*," and moreover takes from it something rather different from what the standard reading might suggest. Arendt does not simply argue that planned economic control contradicts all democratic or local power, nor does she hold that it necessarily swamps all countervailing political organization and aspiration. The very phrasing of communism as electrification *plus* soviets, she observes, implies (surprisingly, from a Marxist like Lenin) that economic or technical institutions may be kept separate from political ones, and that "the problem of poverty is not to be solved through socialization and socialism, but through technical means." (It should be noted, Arendt's reading depends on a very strong stress being given to the word "plus"—and she reads it as simply additive, not aggregative). For Arendt, this holds out the unrealized possibility that the political institutions of local self-representation (the soviets) could have remained sites of democratic realization while standing apart from purely technical institutions that guided electrification. "This is," she writes, "one of the not-infrequent instances when Lenin's gifts as a statesman overruled his Marxist training and ideological convictions."[52]

Arendt ultimately finds in Lenin's formula (which she calls "long forgotten") a *principled* separation between economical and political institutions, between electrification and soviets. Arendt's main argument remains, just as Mehta reads it, one about the fatal effects for true political freedom—and for any possibility of collective thought and action—of the material organization by an interventionist state of goods and goals over which citizens have no control. But this does not exactly lead to a plea for more local control over technical institutions and infrastructures—such as we might hear in present-day advocacy of "participatory development," for instance. Rather, she seems to rest her hopes for a richer political life on a more stringent separation between—or perhaps, a more careful configuration of—technical-economic and political institutions. Let us finish this thought, as it were, by bringing it back to the Indian case.

In the Constituent Assembly the freedom demanded for India and Indians was frequently phrased in material terms, as a function of the technological goods and benefits thought to flow from centralized, coordinated, citywide or nationwide institutions. Yet, still, the politics of Indian constitutionalism and its avowed goals of social reform through "the application of technology" (to borrow Chatterjee's neat phrase) themselves comprised a moral and symbolic project that invited participation in multiple other registers. The material institutions and rational procedures of the central state were, as often as they were imposed from above, also refracted and reflected in poetic projects of self-making.

This last point opens up a level of political possibility different from but congruent with the obscured and historically repressed democratic alternative to totalized statism that Arendt finds in Lenin's formula for communism. Following the lead of Arendt's reading of Lenin's formula, we might seek hidden potentials for autonomous action and for novel solidarities within the very constitution—symbolic and meaningful as much as material and instrumental—of statist techno-politics. The long-term historical results of the independent Indian state's investments in planning, heavy industry, and central economic institutions, as discerned by a critical and retrospective intelligence, cannot entirely efface the rich array of meanings and collective practices to which these very institutions were, and in some ways remain, open. In local encounters with massive technological institutions, we may find hopes for participation, the embrace of technological promises, and even ongoing institutional maintenance of countervailing bonds of local political belonging. Together, these form the cultural sites and the cultural *side* of a dialectical process of state formation. State power must always work, in some regard, through such sites in order to secure and expand its material powers, and they remain arenas of practice and imagination where relative freedom can be exercised—if sometimes only the freedom to dominate others (a critical caution I will explore in the next chapter). There may be few revolutionary soviets on view in this Indian history, but "electrification *plus*" remains a potent formula for critical reflection on the course of Indian state formation.

The lesson to be taken from always-various local encounters with massive technological institutions and infrastructures has been extended and exemplified ethnographically by Jonathan Parry. He has traced, in fieldwork in the 1990s and 2000s among communities around grand infrastructure projects such as the Soviet-built steel plant in Bhilai (in central India), the local, vernacular adoption and novel political use of modernist promises of equal citizenship, technological power, and industrial development.[53] Let me offer, however, a different sighting—also from Bhilai but more contemporary with independence and Nehruvian state building—of the meaningful embrace and symbolic extension of material, technological installations in postindependence India, drawing from the contemporary observations of the journalist and essayist Ved Mehta. Mehta traveled to Bhilai for the ceremonial inauguration of the brand-new steel plant there in 1959, and provides us with a wonderful description of the local embrace of modernist promises and technological devices as emblems of India's collective self-realization.

"One afternoon in February of 1959," Mehta tells us, "several hundred administrators, engineers, designers, and laborers who had been working on the construction of a new steel plant in Bhilai, a dusty, backward village in the heart of India, gathered by the blast furnace, with their wives and children, to witness the smelting of the first iron."[54] The initial smelting was successful, and the assembled crowds cheered the "birth of liquid gold" that ran from the furnace. The hope and the drama of this project were some years later memorialized in a volume of poetry published by the steel plant (in English, with some contributions translated from Indian languages). In the verses of the technicians and workers who were there that day, the inauguration of the plant becomes the occasion for songs of celebration. "Let this garland [of molten iron] glitter around the neck of our mother country," one poet writes. Another provides a couplet of praise:

> Lo! The Blast Furnace is alight
> Pouring molten iron, oh, WHAT delight.[55]

Mehta comments, "The poetic wealth lies not so much in the craftsmanship of the poets, who are, after all, amateurs, as in the quality of their emotional response to the steel plant."[56] This "emotional response" to the steel plant is also an appropriation and a transformation of the very meanings of self-sufficiency, autonomy, and collective becoming with which it—and industrial plants like it elsewhere in India—were ornamented by the politics of the developmentalist state. These appropriations are by no means always as felicitous, or hopeful, or direct as these poems are. But they are no less active and meaningful for being iterations of a high-state politics and charismatic notions of development centered elsewhere, articulated and defined in the English-medium modernity of Nehru but, here, appropriated and transformed in the service of a local project of participation. Such appropriation could perhaps do little, immediately, to shift the "moving ratio" of dependence and autonomy that gives such projects so much of their political importance, but the effort was made, and the feeling of belonging produced—however artificially fostered or ideologically overdetermined—was deep and real (as Parry's later ethnography bears out).

With this example we can see that the political effects—and political affects—of technological symbols do not depend on the intentions or authenticity of the actors who make and circulate political meanings, nor on the strict Arendtian criterion that material and social questions remain separate from political participation and from the reach of political coercion. Rather, what matters is how abstractions integral to political discourse—whatever they may be, including technological imaginations of prosperity, notions of freedom, or relatively novel symbols of identity—are embraced by political actors, turned into elements of a shared political consciousness, and incorporated into life-projects, thus helping to organize changing understandings of time and community.[57]

The idea of central, rational, planned control over the processes of the Indian economy was also, self-consciously for those elites who promoted this as India's pathway to true autonomy, a political effort to produce new forms of self-representation and self-reliance within the nation, to "improve the sense of citizenship" (in Gadgil's words in the debate over the terms of the Electricity Act). Put differently, it was a way to ask and answer the question of the solidarities that would shape the future Indian state, and how they would become the basis for both new governmental power and its democratic legitimacy. The command over the technological and economic powers of the nation that was pursued in postindependence politics was also, integrally, a moral claim on the collective body of the nation, and an attempt to project a future of collective freedom—albeit a kind of freedom bounded by the technological, energetic terms in which it was installed (and of course, this caveat is very much part of the historical point).[58]

This politics was constituted in debates and urgent attempts to resolve contemporary problems, in tense interactions and sometimes contradictory acts of balancing goals of progress and means of participation; throughout, the proprieties of democratic participation were just as much in play as modernist imperatives of progress (as we see with Gupta's objection to central rule in Delhi—an objection that was, however, still phrased infrastructurally and technologically). In short, symbolic devices of (diversely scaled) collectivity and commonality were created through technological installations and rational institutions, at the same time as these became besetting instruments of prospective control over populations. Nehru consistently made his advocacy of central legislative control over—and even direct provision of—public goods not just a matter of necessity, of replacing politics with provision, but also a collective exhortation and appeal. Abstract as the forces impelling world history appear to be in his reflections upon it and India's place in it, the collective task that he articulated was for Indians to work together to seize and shape those forces. As he put it in his book *The Discovery of India*, "It is not lack of power that we suffer from but a misuse of the power we possess or not a proper application of it. Science gives power but remains impersonal."[59]

The political scientist Srirupa Roy has written that much commentary on postcolonial nation-state building in India has missed the fact that the "needs discourse"—the discourse of backwardness, lack, and insufficiency of the sort exemplified by Nehru and so acutely criticized by Uday Singh Mehta—is routinely coupled rhetorically with the "state's 'need of assistance' and the call for people to 'help the state.' "[60] This interpretive point might help us further understand Gadgil's hope that the Electricity Act might help produce a stronger sense of citizenship. This circuit of power is relational and reflexive—the state will build pedagogical institutions, but the citizenry must participate, too, to produce the state's power and reality. A new collective body was addressed and summoned by Nehru's account of India's position amid a metahistory of forces and ineluctable eras. For the historical circuit to be completed, this abstract, collective body had to take on life and reality—become "personal"—in the dreams and hopes of citizens, and be realized through their own ritual participation in the technological body of the state.

4

Broadcast Mantras

The basic content of an epoch and its unobserved impulses
reciprocally illuminate each other.

—Siegfried Kracauer, *The Mass Ornament*

The novelist V. S. Naipaul traveled in India during the emergency—the
period of direct rule under Prime Minister Indira Gandhi from 1975 to
1977—reprising a role he had taken up in his earlier Indian travelogues as
a kind of civilizational goad, steely chronicler of the discomforts and failures visited upon him as a quasi-outsider to his ancestral land. This time,
however, repeated technological failures and touristic inconveniences like
those Naipaul had retailed to the reading public also served as grist for
political reflection on India's rulers, great and small, touching on themes
of autocracy and the spirit of development.[1] On the evidence of his writing
about India, Naipaul can certainly be classed among those "disenchanted
internationalists" of whom Clifford Geertz said that no amount of countervailing argument or experience could shake their convictions about "the
vanity of popular government and the inevitability of autocracy in underdeveloped nations."[2] But Naipaul's journeys in Indira Gandhi's India, perhaps because of his insistence on his own disenchantment, also provide
some revealing insights into local practices of power.

Naipaul visited a relatively prosperous and well-connected village in southern Maharashtra, where he found himself the guest of a village chief, or "the Patel," as Naipaul insists on referring to him. The Patel spoke the language of bureaucracy and state, of universal knowledge and central control—although he himself was little more than a customary, petty despot, and barely more prosperous than the villagers he lorded over with the signs of his special modernity and privilege:

> There was [a] fluorescent tube, slightly askew and in a tangle of cord, in the vestibule [of the Patel's house]; it couldn't be missed. The government had brought electricity to the village five years before, the Patel said; and he thought that forty percent of the village now had electricity. It was interesting that he too had adopted the official habit of speaking in percentages rather than in old-fashioned numbers. But the figure he gave seemed high, because the connection charge was 275 rupees, over twenty-seven dollars, twice a labourer's monthly wage, and electricity was as expensive as in London. . . . The fluorescent tube, like the shining blue sofa for visitors, was only a garnish, a modern extra.[3]

Sixty percent of the village was without electricity, Naipaul observes, while the landlord displayed his modern extras as part of his more total dominance, his "special kind of authority, an authority that to him and the people around him was more real, and less phantasmal, than the authority of outsiders from the city."[4]

Naipaul draws too definite a line of demarcation between the Patel's authority, as *patel* or village headman, and the technical symbols that in Naipaul's description remain fundamentally external to and merely decorative of preexisting relations of dominance. Far from the Patel standing alone in his traditional authority, the "authority of outsiders" is in fact directly at play here as part of the Patel's prestige, and is evident in his stately display of (and apparent monopolization of) access to electric power. The Patel's relations to the state government and its electricity board are at least as much at issue here as his political dominance over the villagers. Being among the "forty percent" of villagers connected to electricity is not only a sign of traditional dominance; it is also part of a modern and technological distribution of power. The symbolic and material value of the electric light cannot be neatly distinguished from each other; its blue glow makes visible and emblematizes the headman's connection to the developmental agencies of the central state, his ability to draw on their power to buttress his local dominance. It is, one could say, the totem of his dominant caste, the status sign that addresses the rest of the village and mediates between his "customary" dominance and his governmental status, modernizing the former in terms of the latter.

Just such appearances of electrical devices—tube lights, radios, cinema, as well as the energy-harnessing machinery of the modern industrial order—amid the putatively traditional political processes of "village India" were not only of interest to disenchanted internationalist observers. The new presence of technology as a sign and a tool of nationwide political transformations was a major ethnographic finding as well as a topic of explicit theoretical reflection in the first generation of professional anthropological research in postindependence India. These sightings of technologically transformed political and social relations, as they are preserved in the ethnographic record, provide an underexploited archive of insights for a broader historical anthropology of technology and politics in India. Moreover, they provide a pathway toward another account of the meaning and power conveyed by displays of electricity like the Patel's tube light, one that is more receptive than Naipaul was to the powers of enchantment and the aspirations of participation integral to India's postcolonial history.

As discussed in chapter 3, technological relations were fostered and legislated for by the postindependence Indian state as a means of national integration and in service of what Nehru and UNESCO alike called the "emotional integration" of the polity (see the epigraph to this section). As Nehru famously said at the opening of the Bhakra Dam in 1963, such infrastructure projects would be the "new temples of resurgent India." But, contemporary development-studies scholars (inadvertently echoing Naipaul's cynical insights) have stressed, "The mass of the people were the objects not the authors of modern development," and "not much is known about the meaning of the new 'temples' to them."[5] In order to discern some of the meaning that electrification could have in "village India," and the kinds of participation in the project of a "new India" that were demanded of (at least) rural elites, in this chapter I return to the ethnographic debates of the 1950s for further sightings of the era's electrically assisted transformations of everyday life and how the new symbols of India's modernity were deployed in local-level practices of political power.

The first generation of fully credentialed ethnographers—of diverse nationalities—to do fieldwork in India could not ignore the changing world of technology and belonging that they found around them in independent India. Because of this, the ethnographic scholarship that was created "in the shadow of empire and the light of nation" (in the apposite words of Saurabh Dube) still has great power to teach us about India's then-emerging democratic publics and politics.[6] There is some truth, of course, to the stereotype that this generation of anthropologists was purely focused on caste, particularistic community, and village life, to the exclusion of "technology, politics, and law," as André Béteille put it in a 1991 review of the state of sociology in and on India.[7] However, on revisiting the ethnography of M. N. Srinivas, for instance, or the theoretical debates between Americans McKim Marriott and Milton Singer on the one hand and the Anglo-French duo of David Pocock and Louis Dumont on the other, we find myriad considerations of "time and the city."[8] As Diane Mines and Nicholas Yazgi have argued, the intellectual creativity of this body of work was generated by the very "tension created by bounding the village off as a closed unit for the sake of analysis, while observing the connections that existed everywhere."[9] Moreover, in this era when sociological "modernization theory" was first being developed—with its notions about how means-oriented capitalist rationality would either transform Indian religious traditions in its own image or would encounter stern religious and cultural barriers to its expansion—ethnographers on the ground found their conceptual resources challenged by the actual patterns of technological use and the repeated and sophisticated deployment of deeply rooted religious and cultural resources to meet contemporary economic and political challenges.[10]

Village Ethnography and Technological Change

Some thirty years before Naipaul's cynical interpretation of the tube light on the village headman's verandah, the anthropologist M. N. Srinivas was witness to similar processes of technological "modernization" and traditional dominance, in a South Indian village near Mysore that he called Rampura.[11] Srinivas had chosen Rampura as a field site precisely because it was not yet electrified, and to the young anthropologist coming out from

the urban center of Mysore, it therefore presented a picture of pristine village life. He soon discovered, however, that village and city, tradition and
modernity, were not so easily cabined apart; Rampura was on the direct
bus line to Mysore, and the traffic of ideas and techniques between the two
was constant.

Srinivas is best known for his analysis of village political economy in
Rampura, and he introduced the term "dominant caste" into the South
Asian sociological lexicon.[12] He sought to understand what social, ritual,
and political forces supported the control exerted by members of one caste
over the economic and religious life of the village. The leading men in
Rampura were all members of the same middle-status *okkaliga* (or peasant)
caste, and as members of this caste they owned most of the land, organized
village labor, and lavishly patronized the local temples. The paradox was
that the *okkaliga* caste had no particular religious or ritual precedence; these
men were neither Brahmans (as Srinivas himself was), nor traditional "rulers." In fact, to the discomfort of this urban, democratic anthropologist, his
wealthy village hosts reminded him of his personal ritual superiority over
them in all sorts of small and large ways. Yet the caste as a whole, owing
to the wealth and political power of a few members, took the leading roles
in the religious life of the locality as patrons and donors. Meanwhile, individual caste members used their traditional position and dominance to
accumulate capital, build modern businesses, and exert political influence
in nearby Mysore and beyond—for example, by 1952 one of Rampura's
landowning families also owned and operated a cinema in Mysore.[13]

In his many rich writings from the mid-1950s onward, Srinivas particularly liked to adduce ethnographic evidence of highly self-conscious
motives for technological improvement in Rampura. Here is one instance
that he returned to in print more than once: "In March of 1948, the peasants of Rampura requested a visiting minister of the State of Mysore for
the loan of a bull-dozer (to level land) and a tractor; for electricity, for a
middle school; and for better health facilities. They wanted electricity as
much for lights and radios as for power."[14] Such ethnographic observations allowed Srinivas to challenge the modernization-theory thesis (itself
based on an overly narrow reading of Max Weber's arguments about the
religious values that motivated capitalism) that said India's development
would be limited by its "traditional" spirituality and investment in ritual.
He was able to show, rather, that ritual performances of status and deeply

religious motivations were not a hindrance but integral to the local embrace of modern devices and economic rationalities.

Srinivas recalls, in his retrospective and longitudinal ethnography *The Remembered Village*, the rise of one local entrepreneur who vigorously worked his connections with state-level politicians and expanded his business enterprises into Mysore and beyond as he gathered influence and capital. Srinivas observed him one morning setting off for a state-sponsored Hindu ritual in Bangalore, the bureaucratic capital city, sometime in 1956. "He was attired in freshly laundered *khadi* and his white Gandhi cap was placed at a jaunty angle. He was about to get into his new, gleaming Ambassador car to go to Bangalore. Mr. Hanumanthaiya, the Chief Minister, was celebrating Satyanarayana Puja, and my friend had been invited. I could not help then thinking of Max Weber who had stressed the other-worldly, 'soteriological' character of Hinduism."[15]

Here Srinivas is pointing out that attending important religious festivals with ministers of state, and arriving in one's gleaming, brand-new, chauffeured Ambassador, can only tendentiously be taken as a tradition-oriented activity (in the terms then used in sociological debate), especially if the religious ritual is a prime occasion of political performance and what we now call "networking." The reverse point equally applies: lobbying a visiting minister for lights and radio in your village or opening a cinema in Mysore is more than simply purposive, rational action. It aims at status exaltation too, by appropriating the signs of progress into local projects, and constitutes a form of participation in a modern, yet still ritual, sphere of transcendent glory.

Radio-Mediated *Rakhi*

Srinivas was not alone in being perplexed and enticed by the perdurance and the flexibility of traditional patterns of political ritual in "village India," as it experienced the massive changes being wrought by independence and by the advent of modern governmental and technological powers. The postwar, postindependence period was also when American and French anthropologists "discovered" India and sought to integrate its complex civilization into their ethnological theorizing. India-based scholars like Srinivas—trained by A. R. Radcliffe-Brown in Oxford—undertook this scholarly task alongside foreign anthropologists, each group shaped by distinct intellectual climates and agendas. However, as Srinivas's research problems suggest, all the ethnographers of this era had to grapple with ongoing political and economic transformations, not despite but because of their focus on caste and village political economy.

Further evidence of new forms of technological honor and ritual distinction, and explicit theoretical debate over the cultural transformations associated with modern technology, can be found in two field-defining works of the 1950s anthropology of India: the collection of essays *Village India*, published by scholars associated with the University of Chicago in 1955, and the journal *Contributions to Indian Sociology*, founded in 1957 at Oxford by David Pocock and Louis Dumont. Despite the aura of agenda setting and school founding that later developed around these distinct projects, the initial publications stand out for their tentative and provisional formulation of the problems confronting any anthropological approach to Indian society. Even the form of publication was relatively informal—an occasional publication of the American Anthropological Association and a periodical mostly authored by its editors, respectively. Each set of essays gained historical weight by later republication and canonization within the discipline, but these essays were at first exploratory efforts, striking out into a very new field for anthropologists—the study of technological and political change in a historically rich civilization.

Village India was the product of a conference held at the University of Chicago in 1954. It collected the ethnographic work of a disparate group of young researchers under the aegis of Robert Redfield's Ford Foundation–funded project in the Comparative Study of Civilizations. Redfield sought to systematize and extend a model of modernization as cultural change that he had earlier developed in studying rural Mexico (not incidentally, during a period of massive institutional transformation in the postrevolutionary Mexican state). Although the papers presented at the conference were theoretically and ethnographically diverse, ranging from economic research to case studies in the "culture and personality" tradition, the overall tenor of the conference seems to have been set by Redfield's ideas about the process of "westernization."

By contrast to this well-funded comparative program, with its loose but coordinated set of theoretical interests, the first issue of *Contributions to Indian Sociology* was a scrappy polemic—Dumont and Pocock sought to define a notion of social anthropology that would allow rigorous sociological comparison between separate and civilizationally distinct entities in East and West. The first issue of *Contributions* included a long, critical review of *Village India*. But let us start with the evidence of technical novelty and social change that the latter volume presented.

McKim Marriott's contribution to the *Village India* volume establishes the ethnographic ground for the subsequent exchange between these two groups of scholars. Moreover, his essay is the most conscientiously focused on the theoretical questions that guided the *Village India* conference and volume: To what extent can a village be a site for studying an ancient, rich, textually deep civilization? Is a village an autonomous social and political unit, an isolated field of relatively static social relations, or only a subordinated part of a larger, historically mobile whole? These questions reflect the theoretical preoccupations of Robert Redfield's concern with social change, but they also rehearse a larger geopolitical mapping of the world into "provincial" versus "metropolitan" areas, rural versus urban lifestyles, and traditional life versus a universal, urban-centered modernity. Such binaries emanated from and supported a progressive, liberal worldview in which ongoing socio-technical change was both desirable and inevitable, and was converging on a single, capitalist pattern (albeit with concomitant "national" variation).[16]

Marriott begins his paper with the broad claim that he will "resolve" the contradictory pulls in the nascent anthropology of India between village studies, with its orientation to local patterns and face-to-face communities, and civilizational learning, with its sophisticated philological and historical demands. He promises to do so ethnographically, by focusing on processes of "communication" between interacting levels of knowledge and practice. He rejects approaches that would divide Indian culture into "horizontal segments" with peasants on the bottom, practicing purely folk traditions, and literati on the top, standardizing and commenting upon textual sources. He further rejects any static binary that would oppose "traditional culture" to "modern" and state-centric forms of life. Rather, Marriott adopts Redfield's vocabulary of "primary" and "secondary" processes of change, with the first being the tradition-oriented, "orthogenetic," ongoing historicity of indigenous culture, and the second being those "heterogenetic," change-producing forces that come at Indian civilization from outside its own borders and traditional terms of thought.[17] It is important to stress again that this is not a binary opposition between "static" traditions and "changeful" modernity—both primary and secondary, orthogenetic and heterogenetic processes are equally innovative, in their own way, but at different metabolic rates. The distinction between primary and secondary cultural processes as Marriott presents it does, however, imply that it is possible to sort out "native" versus "alien" change-producing factors, and that the former will work more in the direction of stability and social reproduction while the latter will tend toward rapid, disorienting technological change.

To give ethnographic specificity to these abstractions, Marriott dwells upon the ways in which a local harvest ritual in his field site of Kishan Garhi is being integrated into more standardized forms of ritual as it becomes associated with the popular Hindu festival of *Raksha Bandhan* or *rakhi* (or, as Marriott calls it, "Charm Tying"). Marriott tells us *rakhi* is typically an occasion when sisters and brothers exchange ritual gifts—when older, married sisters return to the village to tie small thread charms to the wrists of their brothers, and receive small gifts of money from their natal lineage in exchange. This way of celebrating *rakhi*, however, is a fairly recent innovation in Kishan Garhi, standardizing previous harvest festivals, and introducing mass-produced items of Hindu devotion into what was once a purely local affair.

After describing the "orthogenetic" changes that brought Kishan Garhi's local festival into line with Sanskritic models, Marriott tells us, "A further, secondary transformation in the festival . . . is beginning to be evident." Specifically, "the thread charms . . . are now factory made in more attractive forms and are hawked in the village by a local caste group of *Jogis*. A few sisters in Kishan Garhi have taken to tying these heterogenetic charms . . . onto their brothers' wrists. The new string charms are also more convenient for mailing in letters to distant, city-dwelling brothers whom sisters cannot visit on the auspicious day."[18] He continues this exploration of the heterogenetic transformation of a Hindu ritual in a footnote, where he says that another conference participant, Alan Beals, "reports, furthermore, that brothers in the electrified village of Namhalli near Bangalore tuned into All-India Radio in order to receive the time signal at the astrologically exact moment, and then tied such charms to their own wrists, with an accompaniment of broadcast Sanskrit *mantras*."

The analysis stops here, for Marriott has made his point. Industrial mass production, the radio, and electricity are parts of an alien cultural complex, which comes in and alters ritual exchanges—extending, standardizing, and ultimately transforming the very meaning of everyday life as it becomes part of a modern, industrial, urban, national whole. In this example it is almost as if Marriott imagines a complex set of gearing mechanisms, by which local processes are sequentially linked up to orthogenetic, national religious institutions via ritual mediators such as priests and Sanskritic popular culture and then to universal capitalist processes and value orientations, via another set of technological mediators. With the latter, it is as if the whole moral (or immoral) code of capitalism were conveyed along radio waves and copper wires, and automatically adopted by village-level actors when at last they come into contact with manufactured ritual articles and technological means of communication.

As should be evident, this was a highly synthesized example, buttressing the Redfieldian model of change with disconnected field observations. Yet as a bit of ethnographic folklore, Beals's reported evidence of radio-mediated *rakhi* from Namhalli was sufficiently attractive to the other contributors to the volume that it was cited more than once (as far as I can ascertain, Beals never published this observation in his own ethnographies, and it does not appear in his article in *Village India*). David Mandelbaum, in his article in *Village India*, notes that the "ways of the new tradition," which he defines as new institutions of the government, staffed by "medical men, agricultural experts, civil servants," have "penetrated deeply into village culture" in the form of "buses and . . . electric lights." However, Mandelbaum's interpretation of the example from Beals is notably more optimistic about the survival of Indian cultural forms (while also being less systematic) than Marriott's:

> In some ways, an easy adjustment is coming about of old belief and new technique. One example, cited from Namhalli in Mysore, is the use of a radio broadcast time signal at the *astrologically propitious* moment for tying on charms: with the time signal there are broadcast the proper Sanskrit verses. This is a small, if vivid, instance |of adjustment between old belief and new technique|. Radio and other devices of the new complex enter at other levels, as when some villagers in Bombay |province| are said to follow the broadcast price quotations as a guide to the *economically propitious* moment for selling their crop |emphasis added|.[19]

Perhaps unintentionally, Mandelbaum plays here on two different senses of propitiousness, treating astrological and economic calculation as equally expressions of one coherent cultural complex. But this is a fugitive moment of truly comparative thinking in a volume much more concerned with identifying, labeling, and opposing tokens of "old" and "new" integrative processes, fixing each as distinctly religious or secular, primary or secondary.

Technological Change or Structural Transformation?

In their long, programmatic review of *Village India* in the first issue of *Contributions*, Dumont and Pocock argue against this conception of cultural change—which treats change as if it were simply provoked by circulating objects that could be defined, once and for all, by the authenticity or alienness of their origins. They point out, "If we look at the elements or contents of behaviour we are likely to be confused by paradox, but if we look below this behaviour we see that [an underlying, coherent cultural] structure is preserved." Indeed, they write, "an example may be taken from Marriott's article," and go on to cite in detail the example of the manufactured *rakhi* charms and the broadcast mantras. As Dumont and Pocock put it, Marriott "reports that nowadays for the annual ceremony when sisters tie charms on their brothers' wrists, instead of the charms made formally [*sic*; formerly?] by the Brahmans, factory made charms are used and these are 'hawked in the village by a caste group of Jogis. . . . The new string charms are also more convenient for mailing in letters to distant, city-dwelling brothers whom sisters cannot visit on the auspicious day.' "[20]

"Such changes of behaviour," Dumont and Pocock argue, are not "structurally significant," though they may represent a "cultural levelling" of relations previously more markedly opposed and complementary. The "new facts" mentioned by "modern observers" in India, they write, "the spread of western ideas in traditional guise and the spread of traditional ideas by modern methods," such as the coordination of astrologically propitious ritual actions by broadcast signal, "will undoubtedly produce structural change in the course of time, but they will only be understood by structural analysis, not by the proliferation of descriptive terminology."[21] To be clear, let me explain what Dumont and Pocock meant by "structural" significance. For them, structures were social facts that existed beneath the level of ethnographic observation, and which became evident only after careful comparison of multiple observations and analysis that revealed the logic guiding behavior and making it socially significant (but not necessarily conscious) action. They insisted that the spread of "western" ideas was observable and common, but not *structural*. Wage relations in rural employment, political practices associated with democracy and the state, even the things mentioned by Mandelbaum such as new technical experts and schoolteachers in villages, were part of everyday life but did not affect the structure of values, the religiously shaped understanding of the kinds of beings that existed in this world, by which people acted. A schoolteacher was never just a schoolteacher—she was also a member of a caste, a participant in ritual exchanges, and a person from a particular village. These aspects were what oriented most people's relations to her. Moreover, terms such as "orthogenetic" do not help us understand these relations; and anyhow, a gift to the schoolteacher may or may not reflect "traditional" practices, but that was not a matter for observation but rather for structural analysis. Novel practices are structurally important only if the identifications, ideas, and concepts with which people interpret and act have themselves been transformed and integrated into a new pattern of social values. This is a higher bar than was set by Redfield's notion of "heterogenetic" change.

That is to say, Dumont and Pocock do not diminish the reality, or the prevalence, of new ideas, institutions, and practices in the India that they knew from their own ethnographic work in Tamil Nadu and Gujarat, respectively. They diverge from the scholars of *Village India*, however, in that they disagree that such observable changes produce deep, long-lasting alterations in the value orientation of Indians. They do not agree that merely material changes in the mundane tokens of ritual expression have any power to affect underlying ritual attitudes in any way. They argue that a truly comparative ethnography would require something more than observations of the use of manufactured articles, radios, and other devices in everyday life. It would require an analysis from the point of view of structure, where the sort of change that Redfield's terminology describes is not at issue, but rather a rarer, more fundamental, structural *transformation* in whole sets of roles and relations is.

In this sense, Dumont and Pocock advocate a perspective perhaps closer to the one Mandelbaum arrives at with his observation about the convergence of different forms of propitiousness—economic and astrological—now that news of both arrives via the radio. These observers imply, ultimately, that the Indian cultural structure, long coordinated by ritual specialists, easily accommodates economic rationality and its experts too—insofar as they are nothing but new specialists in the abstract, with their own prognostications of mysterious movements and risky convergences.

Radio *Rakhi* Redux

Both of these accounts of tradition and modernity within "village India" are, of course, historically bounded in their visions of change and limited by the very terms of their analyses. While the *Village India* volume does not offer much nuanced consideration of the local meaning and value granted to new technological devices and connections—reading them usually as vectors of change that convey with their manufactured provenance all the values of modernity—Dumont and Pocock do not take the ritual embrace of new devices seriously enough as itself a significant source of transformative power. Indeed, they do not seriously attempt to subject Marriott's synthetic example to analysis on their own terms. Despite this, it is ultimately their sense of cultural structure that allows us to imaginatively reconstruct the changes that might be connected with such an innovation as radio-mediated *rakhi*, and how very structural these changes may be.

Rakhi and its local performances in Kishan Garhi were part of a festival in which connections between out-marrying sisters and village-resident brothers were affirmed. In the "traditional" form of this rite, according to Marriott, sisters exchanged with their brothers to ensure their ability to have recourse—at a crisis, or during childbearing—to their natal village and their relatives there even after leaving for their husband's home. For their part, brothers engaging in these exchanges affirmed the otherwise hard-to-discern moral solidarity of the natal family, even after their sister's marriage. At the time that Marriott observed it, this reciprocal exchange appears to have been in the course of its own transformation from a local, face-to-face exchange into a one-way conveyance, by post, of ritual symbols of affection from village-dwelling sisters to their now distant or absent brothers—who must tie the charms themselves and use the radio to tell them when to do this. At the very least, a ritually significant structure of mobility and residence is rearranged here, since *rakhi* no longer features a reciprocal exchange that marks the relation of the brother's lineage to the now absent "daughters of the house." Rather, now brothers move and become the representatives of the village as a whole in the distant city—which is the site and source of material benefits in a way the sister's marital home never was. In modern *rakhi*, technologically mediated and performed with manufactured charms, migrating men are the medium by which the village women interact, vertically, with the cosmopolitan center—the site of radio broadcasts, and the source of technological goods and national solidarity. Meanwhile, the broadcast of Sanskrit verses and radio-mediated synchronization of the charm-tying obviates the actual presence that was required by the earlier ritual complex, substituting a virtual solidarity for an actual one, and an affect of religious duty for what was an embodied relation of care and mutual obligation.

These transformations are neither "purely" religious nor technological, orthogenetic nor heterogenetic, primary nor secondary. They are, however, significant for any understanding of Indian modernity, as it was extended across infrastructures and transformed in the process, as citizens became participants in the wider "new traditions" of the national state. Broadcast mantras become the emblems of a new level of state power, and the means of the integration of villagers and city dwellers alike into a new community of citizens. One may note that abstract, mass-mediated religious affects may be more conducive to ideologization than embodied and reciprocal obligations, and the history of politicized Hinduism in India might demonstrate this point.[22] Yet that wider observation aside, the point here is that a new technological complex, materially transforming time and space, intervenes tangibly upon the ritual practices of "village India."

McKim Marriott was not as wrongheaded as Dumont and Pocock implied he was, when he saw the circulation of mass-produced charms and radio broadcast mantras as an ethnographically important part of the wider changes under way in the village where he studied. To be sure, it takes a more structural analysis than he provided to reveal the social transformation that was really at stake in his sightings of technology. Yet by the same token this analysis of Marriott's data goes further than Dumont and Pocock would, ultimately revealing mere "changes of behaviour" as fundamental elements in a new political and ritual complex—that of the postcolonial nation-state.

The meaning of this complex to the "mass of Indians" is perhaps not much further illuminated by these theoretical debates among anthropologists, and in that sense this ethnographic archive does not provide a comprehensive answer to the doubts expressed by the development studies scholars discussed earlier.[23] Yet in the sightings of radio-mediated *rakhi*, farmers awaiting news of the economically propitious time to sell, and tube lights ornamenting and securing a headman's village-level dominance, we are still able to glimpse—and maybe get a better view of—the enchantment and the popular appeal of the modern Indian state. At a minimum, these encounters reveal significant processes of technological coordination, integration of villages into higher levels of government, and popular participation in the developmental and propagandistic projects of the state, while showing us—historically—what kinds of things caught the ethnographer's eye in the wider context of independent India. This kind of ethnographic insight is always, in part, a function of what is salient to other people within a given context. And indeed, Srinivas frequently noted the appeal that modern machinery had for his village interlocutors—especially those machines that could be obtained through gifts and grants from state development ministers, who loomed large in the local political landscape. As they appeared in the village, such machines and devices "made an impact on the villagers. . . . Modern technology did indeed perform miracles and human labor appeared pitiful in contrast."[24]

This kind of contrast between miraculous machinery and human labor was much less marked, we might imagine, in the urban capital of New Delhi—where machinery and its energies were both more domesticated. However, even there, attentive observers recorded potent cultural connections between power and belonging, between the technological agency of the state and the scene of political action. Precisely because it is rich with such observations, the Bengali civil servant and writer Asok Mitra's heartfelt appreciation of Delhi's comforts and its civility, published in 1970, provides a nice pendant to Naipaul's more disenchanted observations about the simply ornamental use of electricity by a village despot.

"The problem of Delhi as a capital city today," Mitra writes, "is a problem of integration . . . of leveling up of common amenities [and] bringing about circulations and relations that will bestow on the nine cities [of ancient and modern Delhi] close ties of a common identity." Modern technology and town planning can indeed, Mitra implies, perform miracles, producing "equitable distribution and sharing of wealth and opportunity, greater interdependence and mutual respect."[25] But this is not just a set of recommendations for more top-down infrastructural investment. For him, the real promise held out by the "leveling up of common amenities" is fundamentally a matter of the aesthetics of national belonging, of how Delhi's infrastructural amenities and urban connections might truly transform citizenship and figure the advent of a new morality of urban commonality and interdependence.

Thus, while praising the Delhi government for the efforts it had already made in the tumultuous two decades of independence to house a vast and diverse population, Mitra is quick to note that beyond the intellectual pleasures of a well-ordered city, "there are other charms that more readily strike the eye." With this, he moves into a paean of praise for Delhi's urban, infrastructural landscape. Casting his eye to the horizon beyond the ceremonial core of the city, he describes "the great power house all lit in the middle of a frosty night, still, eerie, so like a painted ship on the edge of the river you could almost hear the sailor's song."[26]

The very means by which Mitra hopes to secure urban and national integration—the circulations and relationships of the techno-political order—are hence transformed, in the writer's imagination, into an image of ethereal brilliance. Technical modernity is invoked not only as development and progress, but also as a magical, miraculous route of access to other worlds. This is a promise of transformation, neither simple ornamentation of existing relations with a "modern extra," nor only of secular change. Actual governmental actors, directly wielding state power, can seldom deliver on such promises, and scarcely dare to articulate them—but the promise remains, in the political meaning forged with new technological capacities and material emblems of belonging throughout this Indian modernity.

Part III

Urban Transformations

Any machinery is an intricate set of similes and metaphoric images
woven through a semiotic fabric (like this one). It is a conspicuous level
of personification, deification, or "demonization" situated, allegory-like,
within, over, or underneath the action itself. Machineries project into
mirrored realms the here and now.

—James A. Boon, *Affinities and Extremes*

5

THE LIFE OF PROPERTY

> Private property has made us so stupid and one-sided that an object is
> only ours when we have it—when it exists for us as *capital*, or when it is
> directly possessed, eaten, drunk, worn, inhabited, etc.,—in short, when it is
> *used* by us. Although private property itself again conceives all these direct
> realizations of possession as *means of life*, and the life which they serve
> as means is the *life of private property*.
>
> —KARL MARX, "Economic and Philosophical Manuscripts," in *Early Writings*

On a pleasant day in the fall of 2005, I entered the lobby of a civic auditorium in the heart of New Delhi to attend an organizational meeting for a
new citywide coalition of activists concerned about rising electricity rates and
the overall management of the recently privatized power grid. The auditorium was close to both the All-India Institute of Medicine—representing
rationality and progress on a national scale—and the Sai Baba Mandir,
where passionate devotion and heartfelt pleas for charity were vividly displayed. This location could not have been more appropriate, for this was
to be my first encounter with an emergent politics of middle-class activism
that was simultaneously technocratic and impassioned.[1]

This activist meeting, which had been publicized via a small notice in
one of the Delhi newspapers, developed out of a mobilization by a loose
middle-class coalition in the summer of 2005 to roll back new electricity
rates set by Delhi's independent regulator. Delhi's formerly state-owned
and vertically integrated utility had been "unbundled" in 2002—broken
into new, separate entities handling supply, transmission, and distribution.

The distribution network was now divided into three "zones," each managed by a private distribution company, or "discom." After this reform and partial privatization, customers had to deal with new institutions delivering their electricity, and the discoms sought to implement new regulations mandating "improved" metering technologies and terms of service. In the summer of 2005, a group operating under the name Campaign Against Power Tariff Hikes successfully lobbied the government and regulator to roll back rate increases. More importantly, key activists in that campaign identified new constituencies developing around power issues within Delhi's neighborhood "Residents' Welfare Associations" (RWAs). Recognizing an opportunity to organize these relatively affluent city residents around the issue of power prices and the outcomes of privatization, a group of formally nonpartisan activists formed a new broader initiative under the umbrella of "United Residents' Joint Action," or URJA, which means "energy" in Hindi.

I arrived at the first meeting of URJA unprepared for the crisscrossing discourses that were voiced there—vague complaints of public corruption, intense assertions of wounded pride and disappointment, and pleas for increased security for private property as well as arguments for strengthening citizenship and civic belonging. I had just started my fieldwork, and given all my reading about urban politics in India, political society, and popular mobilizations against neoliberal reforms, the voices I heard at this meeting were all too familiar in their characterization of a corrupt and inadequate public sphere; yet the tone and manner of the meeting were strange in both the stiff formality of procedure and the barely concealed anger and even rage expressed by these crisply dressed men and women. Two things stood out right at the beginning: first, I was met at the door with suspicious demands to show my qualifications for participation, even though the event had been publically advertised; and second, I was immediately drawn into vigorous conversation about household electricity meters, how they worked and who owned them.

As I arrived, middle-aged men—and a few similarly aged women—milled about, neatly dressed and formal in their interactions. At a table in the lobby, new arrivals were asked to sign in on a form that included spaces for name, address, RWA affiliation, and whether you were an official of your RWA. "What RWA are you from?" I was asked sharply by the men staffing the sign-in table. I somewhat confusedly stammered something about being a researcher from the United States, until someone took me by the elbow and asked me a few questions about my neighborhood in Delhi and its experiences with electric power as he steered me to a seat at the back of the auditorium. I chatted with my guide for a few minutes, and he told me that the main point of the meeting today was to solicit input from representatives of RWAs about their problems with new electronic electricity meters being installed by the private electricity companies. "What kind of meters do you use in the United States?" he asked me with a strange and special urgency. "Electro-mechanical ones or these new electronic ones?" I answered that as far as I knew we still used electro-mechanical meters—thinking of the slowly rotating dials on the electricity meters that had captivated me as a child.

Electricity meters were not, at that point, on my agenda for research. Neither were English-speaking men with degrees from MIT, socializing clubbily while other groups of civil servants conversed in Hindi, before all sat down together in an auditorium to be exhorted, in both languages, to come together to fight for their rights. Yet this meeting—to which I will return—was my entrée into the ethnography of both electricity privatization and the new politics of India's civil society.[2] Despite my initial confusion and uncertainty about what was going on, this meeting clearly represented a local iteration of new global practices of civil society activism, "re-scaling" of the state, and civic governance, all characteristic of the era of neoliberalism (at least as it was then defined in anthropological debates). In cities throughout the world, the late twentieth century saw shifts away from welfarist and collectivist models of local government, and a rapid proliferation of techniques and technologies for reforming old systems of public utility provision. Privatization, decentralization, and participation were governmental watchwords. New forms of accounting, new market mechanisms for the distribution of public goods, and new technical metrics and measurement devices became diagnostic features of neoliberalism for many scholars trying to understand this political dispensation.[3] Under critical scrutiny, however, much of what passed for "reform," promising both greater efficiency and enlarged opportunities for individual freedom and participation in decision making—even if only as a consumer—turned out to increase local autonomy and to expand citizen rights very little, if at all. Scholars found, instead, increased popular dispossession and continued elite dominance, and converged on a critical finding that the vaunted

"rollback" of public power was more often just a reconfiguration or a redistribution, often into formally private hands.[4]

Such accounts of neoliberalism and privatization as part of wider
global political restructuring are by now quite well established. Viewed
through this critical lens, that first meeting of URJA (and the wider "civil
society" mobilizations of which it was a part) provokes a dual set of questions. First, how does electricity privatization and urban reform in Delhi
relate, as a rupture, continuation, or transformation, to the longer-term
processes of Indian state formation and urban politics? And if it does—as
I argue—represent a transformation within, not a rupture with, an ongoing process of state formation, a reorganization of hegemonic powers
over urban processes, how do the middle-class activist mobilizations that
emerged *after* privatization—and did not, as we will see, formally challenge it—stand in relation to these transformations? Second, what role
was played by the interpretive activity that was so important to my interlocutors, when they sought to define what meter replacement and new
terms of connection to the grid meant for their participation in the ongoing
ordering of the city?

To pursue these questions, in this chapter and in this section as a whole,
I explore privatization and its devices as they become elements in a distinctively political consciousness of the present, and how they each were
shaped, interpretively, in relation to earlier urban moments of electrification, ritual, and law. I examine the "improvements" and "reforms" of
privatization as they appear in legislative texts, public meetings, and a court
case over meter replacement that ultimately made its way to the Supreme
Court of India. I recount how the process of privatization emerged out of,
and interacted with, a wider constitutional politics of decentralization and
liberalization, shaped by the legacy of Indian state formation. Finally, I examine how the technicalities—material and legal—of privatized electricity helped foster interpretive claims about the new terms of connection to
the grid and provided an affective basis for the political mobilization and
empowerment of RWAs and their middle-class members as rights-bearing
members of the urban community (albeit a community newly redefined in
highly exclusive and privative terms). In fact, these technicalities provided
the basis for multiple "private constitutional moments," redistributing access, power, and legitimacy and recentering state power, as political status

and rights were claimed in public action and through court cases by the activists of the RWAs and by the discoms.[5]

Other anthropologists studying Delhi's politics and the transformation of the city into a global center and a world city have elucidated the apparent ruptures with past dispensations that neoliberal reforms seemed to both herald and foster, especially focusing on the rising middle classes and their novel discourses of civic order. This scholarship has dwelt on the consumer-oriented "aesthetic" politics of the new middle classes, a politics that has focused on improving urban "amenities" and has had devastating implications for long-term poorer residents of Delhi's core areas—whose homes and sites of work have often been removed to make way for new spaces of consumption and leisure.[6] While building on these analyses, I focus here less on the middle-class discourses of reform and improvement, their material depredations, their limits, and their exclusions, and more on those moments where privatization was, in some way, challenged or rejected by these very classes—who saw themselves as its beneficiaries and, in a way, incumbents—as it imposed new costs and revealed new interdependencies (and simply dependencies). I seek to understand the impetus to middle-class associational life that electricity privatization provided, as its technical impact upon the grid and associated legal reforms spurred new understandings of urban belonging and motivated political organizing in the city's elite neighborhoods. What on the one side is a project of reform and expanded participation, with its distinctive limitations and social boundaries, is also an affective politics where angry responses to failed promises or to novel impositions can grant new meaning to old forms of privilege, recreating them as new and effective claims on status. As the anthropologist William Mazzarella, a specialist in India's public culture, has noted, politics that take as their watchwords "transparency" or "good government" are not to be taken at face value, and not only because they misrepresent that which has to be reformed or "beautified": their power is "as much a matter of affective surges as of carefully calibrated canalization" of needs and desires.[7]

In short, when new electronic meters were installed within Delhi's existing infrastructure, they spurred the formation and articulation of a sense of collective rights and injured legal status among middle-class residents. These meters became, as it were, a collective emblem, and a material site for the production of new understandings of state power and belonging under privatization—and new claims on rights and participation. Moreover, the residents, judges, journalists, and others for whom privatization and its discontents were present and urgent problems, routinely invoked the past in order to understand their present, and frequently offered tropological interpretations in which this or that reform was a fulfillment of some past promise (or its negation). Most importantly, the affective and mobilizational responses to the new political economy of power after privatization replicated, in new idioms and with new incumbents, moral distinctions and technological divisions like those that had structured early electrification in Delhi and which had been central to the workings of colonial governmentality.[8] In this longer view, privatization appears less as an effort to "roll back" the state but rather as an attempt to literally "re-form" it on quite antique terms—especially with the reappearance of status, prestige, and distinction as bases for the distribution of power.

Although I emphasize echoes and repetitions across periods here, the particular accent in which constitutionalism and legality are invoked in the present is a function of, and not independent from, the structural shifts and political imperatives of neoliberalism. A key feature of global projects of reform, and one that was particularly important in Delhi, was a judicialization of politics. The judicialization of politics in Delhi is a separate topic altogether, but the set of shifts that granted judges and regulators decisive authority over electric power in Delhi also enabled the civil-society politics that accompanied and followed privatization. Since privatization and decentralization created new rights and new claims against local political actors, judges found themselves enforcing private contracts as public rights, or compelling state intervention to provide marginalized citizens with access to private markets in the name of their rights. They also ratified the discoms' legal standing, as new authorities over public access to and distributions of power. It is important to stress, then, that as John and Jean Comaroff have observed, neoliberalism comprehends not only economistic rationalities of governance and projects of reform and accountability within states, but also "a hyper-extended, often counter-intuitive deployment of legalities in the social, geographical, political, moral, and material reconstruction of the [local] universe." This is reflected in an "enchanted faith in constitutionalism" that "speaks to something yet deeper: a 'culture of legality' [that] seems to be infusing everyday life almost everywhere, becoming part and parcel of the obsession with order that haunts many nation-states nowadays."[9] On these terms, it is both legitimate and necessary to view Delhi's shifts in urban governance through the analytic lens of neoliberalism, while also noting that its exacting concern with propriety and legality and veritable "obsession with order" were also fundamental features of colonial ritualism. Through fine-grained comparisons across eras we may discover more substantial affinities between colonial and neoliberal forms of governance.

Be that as it may, what matters here is that within a broadly neoliberal culture of legality, as it took shape in Delhi, even well-favored groups still had to deal with their own new positioning within a reformed political economy, and much of what follows shows middle-class activists seeking to understand the shifting constitutional dispensation in which discoms have legal rights and meters impose new obligations, and they themselves are newly subject to market disciplines and yet also solicited to participate in the reform of governance. In chapter 6, to conclude this analysis of the pattern of resident-power under privatization, I will offer an ethnographic account of technical breakdown and repair in the house I lived in during my fieldwork in Delhi, featuring one affluent Delhi resident's nostalgia for the privileges he once had under the state-owned utility. First, however, let us review the legal forms and reforms of the privatization itself, and how they helped spur middle-class associations as new sites of power and authority.

Privatization as Legal Process and Political Drama

In 1998, the administration of Delhi's recently elected (and reform-minded) chief minister Sheila Dikshit took up nationwide incentives toward liberalization of state-owned utilities and started in earnest on the road to privatization, establishing a regulator to oversee the power sector in the state. Delhi's public electricity utility, the Delhi Electricity Supply Undertaking, had already been reorganized by the previous administration in the state government and renamed the Delhi Vidyut Board; but Dikshit's government ultimately pursued a more fundamental reshaping of citizens' relations to electricity. Like earlier efforts elsewhere in the world at privatization and disentanglement of the state from directly economic relationships with citizens, privatization of power in Delhi involved, first, reorganizing the business of electricity around the supposedly distinct technical phases of generation, transmission, and distribution, and second, paving the way for new private actors in the sector by establishing new legal regimes to support new forms of property and instituting new agencies within the state to distribute information and represent divided interests in this transformed economy.

Delhi was exceptional in the wider political economy of power in India and presented favorable conditions for this techno-political reform. Delhi's electricity utility had higher numbers of metered connections and a more compact and efficient distribution infrastructure prior to privatization than most of the other states' famously inefficient electricity boards. Further, operating within a dense urban area politically separate from the adjoining states, Delhi had no large and politically sensitive agricultural sector to subsidize. For all these reasons—and the fact that Delhi offered one of the wealthiest populations of any city in India—its distribution infrastructure offered an appealing target for corporate investment. Still, despite the fact that conditions of power in Delhi were already distinct from those in the rest of India, privatization of Delhi's electricity utility was identified in both law and political culture as an *experiment* in power privatization, a model process that would set a template for reform in other states. In the Delhi Electricity Reform Act of 2000, the Delhi Electricity Regulatory Commission was specifically empowered to act as a local regulator for any state in India, if so directed by the president. Another, less exceptional, aspect of Delhi's reform process was the absence of organized political opposition. In general, as Sunila Kale noted at the time, "there were few countervailing [political] pressures" against state governments' attempts to reform the power sector. "Although public utilities constitute one of India's largest employers, labour unions in the sector had not yet organized effectively to oppose privatization."[10] Indeed, labor unions and even formal political organizations were absent from the contexts where I looked for opposition to the privatization, after it was already a fait accompli. What was present were self-congratulatory accounts of the success and smoothness of the process.

The man who served as the last head of the Vidyut Board said, in an interview after privatization, that "it would be difficult to try to capture . . .

the atmosphere in governance in Delhi" during blackouts and power crises prior to privatization—an experience of "demonstrations, riots . . . and tense, repetitive meetings." This prior condition of politicization made privatization necessary and desirable—as the next stage in a developmental process, a new step on the road to the nation's technical and political maturity. Moreover, privatization was explicitly based on the ideological notion that it would install a new private, commercial relation under which citizens, as consumers, could themselves be held accountable for their own use of power. As the newsmagazine *India Today* put it, in 2003, "there is only one [route to] access to better amenities: turn users into consumers."[11] According to William Mazzarella's neat characterization of urban India's pro-liberalization discourse, the hope—or the political promise—was that the "civil franchise of the *right to choose* triumphs serenely over the messy procedures of representative democracy."[12] The Delhi government publicized its own sanguine certainty (in a press release from 2003) that power-sector reforms "have been received by all sections of the society . . . without any protest, agitation, or labour disturbance." While this was true enough at the time, as it turned out all did not go as planned, and there *was* a collective, political response to privatization.[13] This response, when it came, did not take the form of a comprehensive antiprivatization movement, however, but after privatization ongoing "power woes" did spur activism and public meetings about electricity within the elite civil society that most supported the principle of privatization and formed the constituency for the propaganda from *India Today* and the government's (English-language) press releases.

Part of the problem was that privatization did not go quite as planned from the very first. The auction of the distribution business in Delhi was conducted hastily in the summer of 2002, and early on in the process the international companies that were supposed to provide the reservoirs of capital to restructure Delhi's grid all dropped out. The bid process ultimately produced only two offers, from the large Indian firms Tata and BSES, with *three* distribution zones on the block; both offers were consequently winners, and these corporations formed new discoms as joint ventures with the Delhi state government. Tata established a discom in North Delhi, and BSES set up companies for two distribution zones in South and West Delhi (BSES was purchased by the Indian energy giant Reliance partway through this process, further underscoring the concentration in

the business side of privatization). Later, charges of insiderism and lack
of transparency were raised, and critics highlighted the favorable terms
under which contracts for power distribution in Delhi were awarded; but
these questions about the fairness and transparency of the privatization
process itself emerged only after the fact.[14] Among the first indications
I saw of complaints about corruption in the reform process was an article
in the *Hindu*'s Delhi edition on September 22, 2005, headlined "Money
Power Dictated Privatization"—it is not insignificant that as the campaigns around electricity prices and meters were gathering steam, these
sorts of charges should have emerged in public.

After the initial unbundling and auction of the distribution business,
the further legal and material processes of privatization had to be imposed
on top of a grid that was in many ways already informally privatized, and
in which the most desirable customers actively managed their own access
to electric power. For example, an affluent friend told me that though she
always paid her power bills, her uncle's house was run almost entirely
illegally off the public grid. "None of the air conditioners or TVs in my
uncle's house are even connected to the meter—he had some guy come and
'fix' his meter, and he only pays for enough electricity to run his lights,"
she said. Another friend said that to her knowledge some people in Delhi
just tipped the meter reader, and that was that—no need to worry about
electricity bills!

Such existing "private" points of connection to the grid proliferated between, within, and across households, through various meters, generators,
and, in some cases, direct and illegal hooking into the public grid—and
not only among what Partha Chatterjee has called the "political society" of
the poor (whose distinctly postcolonial form of citizenship is characterized
by more direct, less mediated connections to the institutions of the state).[15]
Any one of these informal connections could coexist in a single installation,
in any household, as the legacy of upgrades, repairs, or work-arounds by
local technicians. In fact, the problem of Delhi's informal grid was less a
matter of rampant criminality and more a question of how social access to
power, and hence a level of *political* participation, had long been distributed
through *technical* means, while technical connections had themselves often
been provided by political influence (what people in Delhi called *pahunch*,
literally "reach"). For decades, intervention in public utility connections
had been a common means of securing access to power, approximating

through these means a universal service that was otherwise out of reach, while the political power of local councilors and informal work-arounds had worked together to shape the reality of the grid. As Philip Oldenburg long ago pointed out, on the basis of a study in Delhi in the late 1960s, public utilities in Delhi "function[ed] well *because* of 'political interference.'"[16]

If privatization was to reform these conditions, further legal and political steps had to be taken. Indeed, the Indian Electricity Act of 2003—passed after Delhi's own electricity reforms had already paved the way for these new provisions—explicitly criminalized irregular connection to the grid, setting forth new categories of dishonest and criminal use of electricity, with new penalties attached.[17] This was one legal move to deal with the already informally privatized grid in the name of more privatization. In a symmetrical reform, the law also "delicensed captive generation," liberalizing the terms on which customers could produce their own power and consume it apart from the public grid. The emerging legal bases for privatization, then, were split between new techniques of legal control, with the new categories of crime, and contrariwise new areas of individual freedom. Alongside these new crimes and property rights, a myriad of new techniques and standards were imposed within the socio-technical system of the grid itself. "Independent" regulation, new performance standards, and "upgraded" metering of all connections were mandated—some by the law itself, some by regulations issued by the Delhi government or the terms of the discoms' contracts.

Privatization was thus a multi-scale and complex series of processes, including state partnership with discoms, legal reforms, and governmental efforts to drive both technical and social changes. Privatization of power also raised expectations; prior even to the impact of meter replacement, discontent was growing in affluent neighborhoods (especially those with existing Residents' Welfare Associations) about the ongoing power problems and continued, routine blackouts.[18] But what emerged in response was not—as some economists suggested at a conference in New Delhi in 2006, as middle-class dissatisfaction became more evident in the city—a straightforward "consumer politics," which would simply evaporate as service improved and prices dropped.[19] It was also dependent on governmental, constitutional reforms that impelled another kind of privatization of power—the recognition of neighborhood associations as a new center of governance.

Bhagidari and Decentralization

At the same time as nationwide moves toward economic liberalization and privatization took shape in the early 1990s, Delhi's government was also set on a new constitutional basis—in part as an outcome of long-term campaigns for "home rule" for the city (early articulations of which we saw in chapter 3). In 1992, the city was made a quasi-state and granted partial legislative autonomy under constitutional reforms that created a new status for Delhi as the "National Capital Territory." Delhi's elevation to statehood, however, happened just as two amendments (the Seventy-Third and Seventy-Fourth Amendments) to the Indian constitution were passed that aimed to encourage decentralization of government functions and specifically to empower village *panchayats* and municipal councils as a "third level" below both central and state governments (collectively, these amendments are referred to as the *panchayati raj* acts; the second applies to urban areas and municipal councils).

Under a formal analysis and in practice, the imperatives of home rule or legislative autonomy for Delhi and of constitutional decentralization conflicted with each other. On the one hand, home rule was itself a kind of centralization of authority, a unification and concentration of executive powers previously shared between the city's different municipal councils and also agencies of the central government. On the other hand, the new Legislative Assembly for Delhi was a state-level government in its own right, within India's federal constitution, and hence the recipient and legitimate incumbent of the lawmaking power previously held by the central government. In practice the municipal councils that had administered Delhi prior to home rule were not rendered defunct, and a paradoxical situation was created: a state government that was the product of democratic decentralization was, at the same time, supposed to empower the "third level" of governance that it had been created to unify and, in a sense, replace. Given the politics of home rule, it was impossible for the new Legislative Assembly in Delhi to contemplate decentralization of its *newly centralized* powers to the municipal councils, and to my knowledge no serious attempts to implement the Seventy-Fourth Amendment were made by the state government prior to 1998.

However, following the election of Dikshit's administration, decentralization and the political demand for more participatory politics within

Delhi could no longer be ignored. The Dikshit government introduced a program in 2000—the same year in which it passed the Delhi Electricity Reform Act—to foster participatory government, under the name of *bhagidari*.[20] This program recognized, but did not provide any new statutory authority for, multiple civil society organizations. Rather than any substantive decentralization, the government under bhagidari merely solicited involvement and consultation on policies and practices from a range of "stakeholders," such as businesses, Residents' Welfare Associations, and NGOs. The name says it all: bhagidari means "partnership," and the word etymologically refers to a share or a stake (*bhag*), revealing its links to good-government discourses of "stakeholder participation." The program's English-language slogan calls it "the Citizen's Partnership in Governance." Unlike other projects of local delegation of municipal functions, like Local Area Management in Mumbai, in Delhi there was little or no defined participation in local administration that came with recognition as a bhagidari partner. With this program, the Delhi state government made efficient use of the existing contradictions between home rule and decentralization, and ultimately also of those that emerged between privatization and empowerment of consumers. The formal contradiction between "home rule" for Delhi on the one hand, and projects of democratic decentralization on the other, is repeated on another level when it comes to the relation between projects of participatory governance and privatization. Bhagidari is meant to reduce the space between the population and the government, to make citizenship practical and engaged by involving citizens in local development and local issues, and to give "residents" (notably, not "citizens") a share in state power (a "share," after all, is just what bhagidari literally promises). The privatization of basic services, meanwhile, works to *increase* the distance between citizens and government by introducing a new actor—the discom, the private provider—between the resident and the very perquisites of participation on which the status of *bhagidar* is predicated.

For the RWAs, bhagidari offered new scope for action. At the time, RWAs were little more than local bodies much like homeowners' associations or condominium boards, common enough in the affluent residential "colonies" of New Delhi and southern Delhi. The groups were limited by terms of membership and techniques of state registration and recognition to neighborhoods that had legal status, formal connections to infrastructures, and a high degree of owner-occupation. Estimates of the reach of RWAs and their class character vary, but in 2005 only a fifth or fewer of Delhi's households had access to representation in a RWA.[21] The very definition of these associations for "residents" is itself a nice piece of ideological sleight of hand, for many of those occupying and working in the neighborhoods—servants, watchmen, vendors, and service workers—are by no means represented by these organizations.

Prior to privatization, RWAs were granted small powers to collect electricity bills, aid in local electricity development, and to read meters on behalf of the Delhi Vidyut Board. Even in 2005 in some neighborhoods, signboards sponsored by the RWA reminded residents to pay their electricity bills. The most salient governance powers exercised by the RWAs throughout the city, however, involved the erection of gates at the entrances to the neighborhoods they represented, and the vigorous assumption of the rights granted to them by the government under bhagidari to control traffic and public space in their local areas.

Anyone who has traveled through the RWA-rich zones of southern Delhi late at night will be aware that there is a secondary geography to these neighborhoods, marked by suddenly locked gates and abandoned watchmen's posts. Each gate is emblazoned with the name of the RWA that erected it. Around parks and other civic amenities within neighborhoods, official-looking signs frequently advise that the park is maintained by a local RWA and sometimes restrict access to local residents. The author of a 2003 editorial in the *Times of India* decried this assault on the very notion of civic space. "The RWAs have in some cases put up big signboards barring entry to outsiders to the public gardens set up by the DDA [Delhi Development Authority—a state agency] and looked after by the MCD [Municipal Corporation of Delhi]. One of the colonies where roads have [been] blocked [to] normal activity is Malviya Nagar where the entry of even . . . [postal] vehicles has been barred to the post office."[22] Indeed, numerous people I consulted observed that the gating of neighborhoods was a relatively recent phenomenon in Delhi—at least on the large scale at which it had been adopted by 2005.

Bhagidari did more than extend a hand in partnership, from the government to citizens (the logo of the program was two interlinked hands); it also created and legitimated—as the expression of local power—new points of control over the movements and circulations that define the city—and reinforced status-rich separations between neighborhoods and even within residential colonies. Indeed, the *Times of India* editorial writer went on to complain that RWAs did not in fact represent a new growth of partnership, an expansion of citizen's powers, but rather a return to types of landownership often described in Indian political discourse as feudal. He asked, "Is this *bhagidari* or *jagirdari*," partnership or landlordism?[23]

This editorial says a great deal with this minor punning comment. With bhagidari, a highly dominating politics began to emerge, establishing moral and community boundaries (often specifically religious) that were, not incidentally, coextensive with new practices of consumption.[24] These arenas of neighborhood politics, especially as they became prominent within the political economy of privatization, provided space for claims of status buttressed by an assertion of authority over the urban processes—of circulation, of material provision, and of security—that have long been the object and purview of state power.

URJA: Energy in Hindi

The first meeting of URJA that I attended, in October 2005, was highly organized and was gaveled to a start more or less on time by a representative from the main sponsoring NGO, People's Action. This quasi-politician (who later became a spokesman for the Bharatiya Janata Party in Delhi) soon turned the stage over to the president of one of the more affluent RWAs, who served as moderator. He stood at the podium and addressed the surprisingly full auditorium about the high bills his residents had been receiving. Called up by the moderator, speakers began testifying about their recent problems with meters and billing. The testimonials and accusations grew fiercer as the morning ran on: a stack of photocopied bills—some running to tens of thousands of rupees—collected from just one colony were presented to the RWA president as evidence for the group to consider. What could possibly explain this explosion of vastly inflated bills, the president asked the room? One speaker charged the Delhi regulator with colluding with the discoms and licensing fraudulent meters with margins of error greater than the 3 percent allowed by law.[25] Someone pointed out that the new meters were manufactured in China, implying that fraud and corruption could be expected from that geopolitical rival. Finally the man

I had spoken to before the meeting announced from the stage that there was a visitor from America in the room, and that I had told him even in the United States they did not use these false electronic meters!

The meters were not the only concern, of course. Theft of power was also a major theme of the presentations. "Look around you," the president asked at one point, "look in your own neighborhood—who is stealing power? In my own neighborhood, even the *mandir* [temple] is hooking power, and the shops in the marketplace have electric lights strung up everywhere!" ("Hooking" means to illegally connect to the overhead wires.) He brought the room's attention to the main point: Whose meter, he asked darkly, might be connected to those public displays of illumination adorning neighborhood temples and marketplaces?

Of course, these concerns about power theft were entirely valid, as were worries about power drawn from illegal connections being charged to legitimate meter owners. But there was something curious—perhaps indicative of a wider consciousness of legality as an idiom of citizenship—in the focus on theft and criminality, and the concomitant emphasis on the meters as themselves "corrupt" and even some kind of insult to India's sovereignty (since they were made in China).

The organizational meeting of URJA was not the first public appearance of the meter issue; but the prominence of metering as a focus and an explanation for high electricity bills seemed to take the NGO sponsors of the event somewhat by surprise. The lead organizer's closing speech somewhat awkwardly called for more rather than less government surveillance of movement and mobility in the city, in order to deal with the problems of plebian criminality on which he wished to focus; although throughout the meeting, RWA representatives had been stressing the burden similar surveillance and control was placing on them when it came from the private discoms in the form of new meters.

Retrospectively, this meeting was a key moment in the emergence of a post-privatization activism and the integration of the RWAs into a larger political transformation, amid the processes of legal and material change in Delhi. Starting with the Campaign Against Power Tariff Hikes, middle-class protests, not against privatization but for *better* privatization and results, expanded and transformed throughout late 2005. One journalist observed that "the protest was a middle-class and upper-middle-class movement and not a 'mass movement' as described by certain sections of the media." Further, "the RWAs are not calling for an end to privatised power distribution. They are asking for enhanced competition through open access and better results."[26] The character of the RWA mobilization as "not a 'mass movement'" but instead a protest against state policies, in fact, tracks closely with larger shifts in Indian political life. Indian feminists, in particular, have noted the replacement in Indian political discourse of the nationalist, collective exhortation to participate in development through ascetic and collective practices with a more particularistic, identitarian logic of empowerment and struggle—situating citizens in opposition to the state and as needing to be "empowered" by NGOs on the basis of their status as women or minorities (or, as here, "residents"). Citizens are, in this logic, asked not to participate *in* the state but to struggle against it, and mobilize their own particularistic need, to secure their rights.[27]

The RWA activism in Delhi, to be sure, developed from consumer discontent with high electricity tariffs and emerged as a movement for transparency and better results. But it drew strength from the promises of "partnership" in governance that bhagidari put forth and built upon logics of state decentralization in order to foster a new practice of politics through the mobilization of novel statuses based both in property ownership and government recognition. Moreover, this activism was not nearly as "civil" as it appeared at the outset. While it had all the appearances of a mediated "public" discourse (in contrast with the direct actions and forms of protest more usually associated with Indian urban politics)—taking up newspaper column inches and being played out in editorials in elite publications, meetings in plush auditoriums, and legal suits over infringements of property rights (as we will see)—more immediate expressions of agitation, passion, and political affect were also evident. The new meters attracted charges of corruption not only because they were associated with high bills but also because they impinged on these residents' management of their own connections to the grid and to other citizens and injured their collective sense of status and political distinction as law-abiding partners in government.

Choose Your Own Electronic Meter

As it turned out, the new electronic meters themselves were in fact the source of the high bills that sparked such debate at the URJA meeting—but not because they were fraudulent or because the discoms were corrupt. Rather, the electronic meters were designed to collect better and more comprehensive data in order to detect theft and tampering. The new meters were meant to be a simple technical fix to the problems of Delhi's under-regulated grid, but these new measurement capacities conflicted with the technical-cum-social conditions of Delhi's grid itself, composed of multiple work-arounds and technical patches.

The new electronic meters were designed to detect fraud and circumvention by measuring the current both going into and coming out of the house. If these two measurements (phase and neutral, respectively) were not equal, it would indicate that power was being drawn into and consumed in the house from some source other than the main metered connection (that is, the meter had been circumvented). At least, that is how the meters were designed to function, the idea being that the power company could simply bill for the higher of the two measurements and avoid being cheated by simple and common tricks of wiring around the meter. However, many affluent neighborhoods in Delhi had previously been wired, efficiently and not at all fraudulently, in such a way that all the houses in a street or block *shared* a "neutral" line. This was possible because the old electro-mechanical meters registered only the "phase" draw (this is of course also what allowed them to be circumvented easily). For those who shared a neutral line with many houses in line before them (or with a temple or marketplace or other urban institution that had illegally tapped into the grid), the new meters would thus register not only their own actual consumption, but also the aggregate neutral load of every electricity user "upstream" of their meter.[28] Combined with the existing infrastructure and with new billing practices, the new electronic meters in Delhi did over-measure and overbill individual consumption, and they had the effect of revealing a mechanical interdependence *within* neighborhoods, one that had not previously been a problem. The activists at the URJA meeting were right—their new meters were registering the consumption of many people in their neighborhood and overbilling them wildly. Whose property was at stake, and how, was the question that had still to be answered.

Coincident with the Campaign Against Power Tariff Hikes and just as the meter issue was gathering force in Delhi, the Delhi Electricity Regulatory Commission, along with the central government's Bureau of Indian Standards, launched a "meter testing drive," while vigorous attempts to were made in the press and in the form of pamphlets included in the power companies' bills to assure the public that the meters met all the highest standards.[29] In most respects, the meters did meet those standards—technically. In the vast majority of cases where some problem was found with the meter readings during one testing drive, the problem was identified as a "shared neutral," or otherwise "faulty" wiring. A meter-awareness website run by the Delhi government attempted to quell complaints about the new meters by shifting responsibility for high bills and erroneous readings from the meter to the wiring of the individual household (despite the fact that it was the wiring between households, not within them, that was at issue). The website noted that "the problem of [shared] neutral wires arises from faulty wiring in the consumer's premises and this has to be rectified by the consumers themselves."[30]

In an interview with the *Hindu* in November 2005, an RWA president expressed his disagreement with the government's interpretation of who was at fault here, and who bore responsibility for fixing the bad wiring. "The regulator has completely lost the confidence of the people and is dancing to the tune of the private discoms," he said. "Why was the consumer not informed that shared neutral can result in fictitious billing? Why should the consumers shell out large sums to rewire their houses as a safeguard against fraudulent meters? Why were the discoms allowed to extort extra money on the pretext of checking [stopping] power theft?" he asked.[31]

The discoms, for their part, explicitly sought to solicit residents' individual participation and investment in fixing these problems, too. In advertisements placed in the newspapers, the discoms extolled the virtues of the new electronic meters and highlighted the improved safety and convenience that they provided, while encouraging consumers to purchase and install one of these new meters. "The advantages of an electronic meter are greater than you can imagine," readers of the *Hindustan Times* were told in one advertisement on July 25, 2005. "Choose Your Own Electronic Meter," a mailer included in bills that summer and fall proclaimed in large orange type in both English and Hindi. "Now the decision to replace your defective meter with the new BSES electronic meter or the one of your choice, rests with you," the ad copy explained. The "choice" offered, however, was limited to the selection of a meter model from an approved list of vendors, and the wording of the mailer did not offer the choice of *not* replacing

Figure 6. "Choose Your Own Electronic Meter," mailer included in BSES bills, 2005.

your presumptively defective electro-mechanical meter (the Hindi mailer referred to customers' existing meters as *kharab*, which means bad or defective but with a stronger moral connotation than the English carries). Instructively, these ads and the advice to consumers distributed with bills also stressed that the new meters were able to measure flows of power beyond what the customer had contracted for—making it easy to consume more power and be billed for it, if you so desired. It was, then, this very socio-technical bargain built into the legal design of privatization, promoting on the social side *increased freedom* to consume and even produce as much power as one wanted, on the technical side *increased surveillance* and control over individual connections to the grid, which meant that massively inflated bills were regularly arriving at householders' doorsteps.

Throughout the meter controversy, the regulator and the discoms tried to keep the focus on the individual consumer's behavior and property, localizing responsibility by focusing on the wiring of homes as the problem, and eliding the contribution of existing neighborhood wiring and the measurement technologies included in the meters themselves. Meanwhile, the residents were caught in an impasse: their bills in many cases really were illegitimately higher, but given the terms on which they organized as "residents" and consumers, and their wider concerns with civic order, they couldn't very well explicitly object to paying for *their own* consumption of electricity. Indeed, the constant reference to bills is itself significant—one common theme was the claim that "we have always paid our bills" and yet were now being stuck with the costs of everyone else's corruption and criminality. The fact that these citizens had bills to brandish, as they did at the URJA meeting, was a sign of their political standing, their status within the wider economy of power in the city, granting legitimacy to their demand that the government stop meter replacement yet continue to discipline unruly consumers elsewhere.[32] In this context, the meters themselves provided a tangible site of legitimate grievance while also—in their novelty and their ubiquity—offering a shared object around which to organize and express discontent. By focusing on the meters and the power to bill and regulate that they represented, the members of the RWAs were able to give collective voice to the newly revealed individual costs of privatization.

New Agencies of Progress

As the controversy over the new electronic meters was taking shape, with URJA publicly calling out the "fraudulent" electronic meters, the regulator's meter-testing drive under way, and BSES advising its consumers to buy their own electronic meter, Suresh Jindal, a resident of an elite neighborhood of Delhi, sued his discom to challenge their replacement of his old, electro-mechanical meter with a new, electronic one. The suit, *Jindal vs. BSES Rajdhani* was rapidly decided in favor of the discom at the Delhi High Court, in December 2005, but it was ultimately appealed all the way to the Supreme Court of India. This case offered a jurisprudential resolution of questions about both the property at stake in privatization, and the propriety of the program of meter replacement, ultimately ratifying the discoms' authority and even their legitimacy by appealing to the technological advance that the new electronic meters represented. The courts that heard this case answered an individual's grievance by reading collective purposes and meanings of national progress into privatization and its devices.

Jindal offered three arguments against being compelled to install or accept a new meter: first, that his original meter, which he owned, was fully functioning; second, that the new, electronic meter was in fact defective, resulting in erroneous and very high charges for electricity; and third, that the discom had no right to replace his property with theirs in the first place. Jindal demanded reinstatement of his old meter and claimed his right to choose what metering technology to use (this was not an unreasonable claim, since customers were simultaneously being encouraged by discoms to choose their own electronic meter).

The High Court judge who heard this case declined to rule on Jindal's claim that his old meter worked better than the new electronic ones, and he based his ruling against Jindal largely on the notion that the electricity companies were simply encouraging him to improve or upgrade his technological property, while implying that this was a collective responsibility of good citizens. The electronic meters, the judge pointed out, could register and store more information about the connection, communicate that information to the discom, and facilitate coordinated management of the grid as a whole—although he acknowledged that meter replacement involved an intrusion of sorts into the residence. As he put it, meters occupy "very little space in our homes. But the advent of electronic meters in Delhi has stirred a hornet's nest. [Their] accuracy, credibility and if I may say, in the eyes of some, even [their] integrity is in doubt." Still, he concluded, "the electronic meter proclaims: believe me, I am smarter than my ancestors."[33]

This "hornet's nest" to which this judge refers is, effectively, the anger and sense of insecurity and doubt about electric power and its progress in Delhi, evident in public politics throughout this period. The judge directly links this buzz of political activity to the advent of electronic meters, but he then personifies the meter, imagining it speaking to the consumer to assure him that progress is at hand. In this decision, then, the meter itself installs the improved and modernized power relationship that privatization had promised. Although the legal reasoning is a bit muddy, this decision further implies that there really could be no reason to reject the new meters since they themselves *articulate progress*. This decision determines the electronic meter to be superior to the old, electro-mechanical version from first principles alone—the superiority of electronic to electro-mechanical meters is assumed and is not based on any technical assessment of the meters at issue or investigation of the wider distribution network in Delhi (in which, in fact, the new meters did not function better than the old ones).

When the issue finally reached the Indian Supreme Court on appeal in 2008, that court agreed with the earlier judgment, but on different grounds. The Supreme Court narrowed and refined the lower court's analysis, reducing the point at law to whether or not the discom, as a successor to the previous state utility company, had the right to demand changes in its customers' meters, and to alter the terms and means by which the customer connected to the grid.[34] The Supreme Court found that "there does not exist any provision in any of the statutes [governing metering and licensing of discoms] which precludes or prohibits the licensee to replace one meter by another." In short, the discom had the right to control metering absolutely; if the state-owned utility had previously allowed people to install their own meters, this did not change the fact that the property interest at stake was not the householder's ownership of the meter itself, but rather the discom's interest in the electricity it delivered to the house and for which the meter helped it recover full value.[35]

This legal ruling was further supported by a justification drawn from the principles of statutory construction, which oddly inserts again the progress or "advance" of science and technology as the criterion of judicial action. The court wrote, "If such a provision [barring meter replacement by the discom] is read into the statute, the same would come in the way of giving effect to the benefits of new technological development. Creative interpretation of the provisions of the statute demands that with the advance in science and technology, the Court should read the provisions of a statute in such a manner as to give effect thereto."[36]

The court's notion of "creative interpretation" is supported by a reference here to a standard handbook of statutory construal, which specifies that the underlying purpose of a law ought to keep pace with technological changes.[37] For example, statutes including reference to telegraphic communication should be read to cover new forms of long-distance communication, too, such as telephony and even e-mail. But the Supreme Court's reasoning goes further (echoing the lower court judgment) to imply, again without evidence, that the electronic meters in themselves can represent *nothing but* an advance in science and technology—any rights-based limitation on a discom's power to alter a customer's meter would "come in the way of giving effect to the benefits of new technological development."

This insistent claim that electronic meters could only represent an improvement (voiced, as we have seen, by the discoms and by the justices at each stage of this case) was directly confronted by Jindal's truthful charge that the electronic meters did not represent an advance, did not function better than the old-style meters—at least, in the conditions of Delhi's grid. But the question of whether a given meter functioned better or not, the court ruled, could be decided only on a case-by-case basis by the discom.

With this decision, the Supreme Court recognized and ratified the discoms' new authority, based in their property interest—not only an authority over the material fabric of the grid, but also over all the devices attached to it, the energy that flowed across it, and ultimately over Delhi's residents themselves, since a discom could compel them to upgrade their wiring by simply pointing out that the new meters were, when viewed in isolation, superior to the old ones.

Put a different way, the Supreme Court decision in this case counters Jindal's plea of his right to choose whatever meter he desires, grounded in a property right, with the discoms' supervening right to decide what meter he will be allowed to use, grounded in the state's delegation of power to the discoms, as the new agencies of progress. This represents a fundamental allocation of sovereign power, one that is achieved not just "by privatization" but also as a matter of interpretation. The court is specifying what the meters do, technically, and how that action gives form to social progress. The powerful effect of this decision cannot simply be dissolved back into a negotiation of new capacities, new powers, and new rights among the plurality of actors and agents with stakes in this controversy, nor does it have only to do with the technicalities of metering. A legal technicality (statutory construction) and a notion of collective progress are also essential here. Ultimately, the court decided who had legitimate power to control metering and indeed all electrical relations in Delhi, and the discoms iterated that decision throughout Delhi in their ongoing promotion of meter replacement.

This judicial finale does not mean that the RWAs' activism was fruitless; with their meetings over high power bills and protests against meter replacement, they highlighted a contradiction in neoliberal state formation, and ultimately perhaps resolved it in their own favor despite the adverse decision of the Supreme Court. With privatization, the private sphere is treated as the locus of a public political agency, and a purely private contractual authority—to meter, to bill, to install—becomes almost a form of public power. While they lost the battle against meter replacement, the RWAs continued to proliferate as a quasi-governmental form and to shape and regulate spatial and political relations within Delhi. RWAs provided just the vehicle for some "residents" to find purchase within this new configuration of private power as representatives of a kind of public authority.

The transfers of power that in one account of privatization might be represented as a succession of delegations from a political center to ever-more *apolitical* technologies and institutions (from the Delhi Vidyut Board to the privatized discoms, or from tamper-prone electro-mechanical to "smarter" electronic meters) in practice required massive public and material effort to rewire the locations and techniques of governmental jurisdiction and construct the legitimacy of new kinds of powers. That effort is what becomes visible as we follow, ethnographically, these interpretive conflicts over the legitimacy, accuracy, and the truthfulness of the new electricity meters.

What is significant, then, about these activist claims about fraudulent meters and judicial interpretations of what they mean for the progress of society is not just a particular conjuncture of economic privatization and political decentralization—congruent, or twinned, political processes in much of the world. Rather, this political and legal controversy reveals how social actors that have no particular legitimacy or even real institutional existence apart from their identification with the "local" and the "private"— on the one hand the Residents' Welfare Associations, and on the other the discoms—gained a new authority and indeed power as public, political agents. The RWAs and the discoms were each able to carve out public, political space for themselves and indeed to effect changes in the very terms of participation in the public sphere—precisely because of their claims to represent legitimacy, progress, and reason on terms, and through symbols, once monopolized by the developmentalist state.

Mechanical Disconnections

Jindal lost his case and could not choose what sort of meter he would install in his house; the discoms have not finally been victorious, either, in claiming a total power over the grid in the name of economic efficiency and technological progress. The Delhi state remains a shareholder in the new privatized discoms, and state regulation retains its force. Meanwhile, what began with middle-class associations and an NGO-led movement for transparency in privatization has broadened once again into a mass politics (in part owing to the creative, charismatic leadership of the Aam Aadmi Party) rather than the affectively saturated but still civic forms of assembly and legal procedure that I have surveyed here.

For my purposes, what is most significant is how decentralization and privatization together involved plural delegations of stately powers to private, and privative, actors, and how these delegations were known, seized, and reworked through key symbols like the meters. Affective surges, provoked by the new impositions and exactions of a quasi-state power, then can interact with interpretive processes—what does a high bill or a new meter mean, politically? Both work in and through the materiality of the meters and residents' concrete concerns with high electricity prices, without necessarily being reducible to either, and constitute new political realities by fostering novel recognitions of collective agency and distributing power and rights.

Ultimately, however, because of the duality of this politics—riven between property rights granted to discoms and participatory rights promised to citizens—we are confronted with a situation in which the promises of decentralization—increased local participation and involvement—are *fulfilled* at the cost of a really increased isolation, fragmentation, and abandonment of large areas of the city to the control of privatized authorities. As has been observed amid neoliberal urban reforms elsewhere, the paired processes of decentralization and privatization represent something of a "technocratic bait-and-switch"; real control and power over parts of the city are granted to some, while at the same time others are situated in a new relation of subordination to novel, and often unaccountable, authorities.[38] It was this contradiction that the members of affluent RWAs registered, with their angry response to the new meters.

The ultimate orientation and final effect of this activism might be best captured by a formula that Marx developed, long ago, representing the final alienation of social being in capitalism from its human sources of power and productivity. Under the rule of private property "an object is only ours when we have it," he wrote; and if "private property . . . conceives all these direct realizations of possession as *means of life*," the only life such concrete, immediate acts of possession serve is "the *life of private property*" itself.[39] The legal provisions of the various electricity reforms, and the promise of privatization, seemed to offer a utopia of private control over electric power, through direct action upon and autonomous control over the electricity grid, heralding a new kind of progress for all of society. But the actual result of privatization was to re-center power (in the discoms) and hence to serve the life of capital.

Yet, a ritual and symbolic politics had to be traversed to reach this result, one that cast into doubt the legitimacy of particular forms of action in public space and raised a set of constitutional questions about the distribution of power and rights. To be clear, these questions and how they were resolved—both in courts of law and through processes of technological assessment and standards setting—were sociologically but not formally constitutional, since the Supreme Court case here was a statutory one. But this narrative is, certainly, full of what Annelise Riles has called "private constitutional moments."[40]

The core symbols of the state and of plenitude, of bureaucratic favor and status, which we have seen being constructed over time throughout the various installations and extensions of electricity in Delhi, were through the activity of the RWAs shifted, refracted, and reflected into new configurations. These new configurations of property and privilege do stand in opposition, in many ways, to the promises of national integration and of "leveling up" of material standards that were part of the symbolic drama of constitutionalism in independent India. Far from symbolizing mutual and reciprocal interdependence—that is, an "organic" solidarity, in Durkheim's terms, predicated on division and difference—the new meters materially revealed only a literally "mechanical" entanglement within neighborhoods and between residents. The mechanical solidarity—a political relation of sameness and unity, and jealous antagonism between opposed groups, rather than a differentiated interdependence—revealed by the new electronic meters as a relation to a "shared neutral" line, was in turn made into a political status of "resident" from which new rights and new claims on power could be articulated but not new partnerships across urban divisions. The meter controversy and bhagidari alike operated to reinforce a singular identity—that of law-abiding "resident"—as a basis for participation and an associated mechanical solidarity as the foundation of effective citizenship.

This is not an isolated instance in contemporary politics of a new form of power being created through the cultivation of local bonds imagined to be those of identity and similarity between individuals, and ultimately wielded against others on behalf of a moral reform of the state or the political process. As the legal anthropologist Carol Greenhouse has pointed out, as a social theory neoliberalism inverts the temporal relation between the poles of mechanical and organic solidarity that Durkheim posited, in which the latter progressively replaces the former through the growth of the social organism. By contrast, Greenhouse writes, "In its valorization of the individual, its preference for markets over rights as the basis for social reform, and its withdrawal of the state from the service sector, neoliberalism overwrites older notions of the public based in organic solidarity with a strong mechanical overlay—as an improvement, or modernization, of more traditional social bonds." Greenhouse goes on to argue that "understanding this inversion is crucial to understanding the nature of the interpretive questions to which neoliberalism gives rise in everyday life, since neoliberal reform reshapes the relationship between society and the state without eliminating what came before."[41]

If neoliberal reforms are justified by their ability to cut through sedimented political dependency (and an atmosphere of governance said, by reformers, to be characterized by "demonstrations, riots" and "labour disturbance") in order to free the individual as an economic, rational, self-interested actor, in the case at hand the political result—as Greenhouse's analysis anticipates—is to strengthen novel forms of mechanical solidarity as political statuses, and hence as the basis for some groups holding power to themselves, and ultimately wielding it over others. However, the new status of "resident," cultivated by bhagidari and the new techniques of surveying and metering electricity together, creates a configuration of rights, privileges, and material wealth that in some ways reproduces the politics of prestige on which colonial forms of urban governmentality so centrally depended for their efficacy and force. Prestige, status, and sumptuary privileges are in play in this contemporary urban politics—and these nonmodern terms thus retain their analytic relevance at the very heart of a highly rationalized, neoliberal biopolitics of surveying, regulating, and micrologically controlling the power consumption of a whole urban populace.[42] Reformers strategically situated RWAs at a new level of democratic legitimacy, and judges recognized discoms as the authorities who would be able to improve technical life. In these interlinked processes, discoms were garlanded with the legitimacy accorded to the progress of science, and "residents" were accorded recognition as all-but-totalizing representatives of the urban polity. Both were thus able to exert on their own behalf the police powers of the state and were granted a kind of sovereignty over their special domains.

Whatever the resistance to meter replacement mounted by these middle-class activists, the privatization of electricity seems to have meshed very well with the new logic of state decentralization and empowerment of the middle classes through partnership. While these middle-class activists complained about the "high-handed methods" of the discoms and resented the loss of privileges over their own property that was both caused and symbolized by the new meters, for their part they organized around a status that was itself based on an informal delegation of state powers to police and control their neighborhoods. This new status was just as privative as the authority accorded to the discoms. The combination of privatization and delegation did not improve participation and collective freedom for the mass of citizens, but rather, paradoxically, shaped new claims of authority, initially within neighborhoods and then on a citywide basis. Middle-class RWAs continued, and still continue, to pursue gating, segregation, and local dominance while cultivating a larger politics of "residence" as a mechanical qualification for civic participation, reinforcing through these sumptuary consumer politics and moralizing aesthetics the more material stratifications of the social order.[43] And on this understanding, we may rightly call the politics pursued by the middle-class RWAs and their present dominance "stately" and identify a deeply structuring ritual dimension inherent in their civic meetings and in their technological politics of electricity connections and electronic meters.

6

A MODEL COLONY

And so far as political theory is concerned, it is there, in exposing the
symbolic dimensions of state power, that the use of attending to decaying
rank, dispersed prerogative, ritualized water control, alien-managed trade,
and exemplary cremation lies. Such study restores our sense of the ordering
force of display, regard, and drama.

—Clifford Geertz, *Negara*

A power failure in the spring of 2006 knocked out power one evening in my Delhi neighborhood. While the power had gone out before for three or four hours, this time it stayed out overnight. By force of circumstance and by dint of good connections, I had rented an apartment in one of the influential and favored neighborhoods in south Delhi. It was close to New Delhi proper and was, locals said, a good "colony," or affluent residential district. Blackouts of any kind were rare in this neighborhood—and outages of more than a few hours were almost unheard of, so my apartment did not have a private generator but only a backup battery system (or "inverter") to keep lights and fans going. In fact, I sometimes regretted that I was so insulated at home from experiences of power outage like those I regularly sat through elsewhere in the city. Even so, on this occasion the outage barely disrupted my routine, and I weathered it with minimal inconvenience.

In the morning, however, when the main power was reconnected, I noticed strange changes in my electrical goods. Some switches and outlets in my apartment worked, while others did not; stranger still, my doorbell rang intermittently, even though there was nobody at the door, and when I fired up the television, the top of the image had an eerie purple cast. The whole kitchen and foyer, around the circuit breakers, inverter, and meter, stank with the acrid smell of burnt ozone. I contacted my landlord, Mr. Mehta—who lived in the main house in the same compound and whose power had also been cut off the night before. After some poking around in the neighborhood, he told me that the outage had been more than just a commonplace fluctuation; a team of workmen digging up the streets had cut a feeder line buried in the alley behind the house, and this had somehow sent a surge through the line that supplied both his house and the small, separate building where my apartment was located.

As we went about figuring out what needed to be done to fix the electrical problems in my apartment, my landlord grumbled about this localized blackout but also complained at more length about the new conditions of power after privatization. In this colony—like many others—responsibility for electricity infrastructure was now disaggregated among private power distributors, the neighborhood improvement association, and—this local outage revealed—also subcontractors and local workmen tasked with various aspects of upgrading the local grid. What this meant was that no one would take overall responsibility for the work going on throughout our neighborhood, nor would any material help be forthcoming to repair the damage it had caused in Mr. Mehta's compound.

A couple of days later, after a long process of calling around and collecting references for various workmen and trying to decide whom to entrust with the work, a *mistry* (technician) was hired to fix the problems in my apartment, and he and his assistant took up residence while they tore out all the old wiring and installed new "electricals." In the meantime, I moved into the main house with Mr. Mehta and his wife.

I spent an afternoon talking with the mistry as he opened the walls of the apartment and pulled out masses of fused, burnt, and melted parti-colored wires—wielding a small screwdriver that lit up whenever it touched live wires. The thin plastic coatings that were supposed to insulate the wires had melted and blackened in the power surge, and the mistry's investigation revealed that much, but not all, of the apartment's wiring was directly connected to the main power supply for the compound, circumventing the rack of circuit breakers that adorned my kitchen. I asked about the catastrophic effect of the surge: Was it that the apartment's wiring itself was badly installed, or was this all just an effect of a large and unexpected spike from the work along the mains? Indicating the melted wiring running behind the circuit breakers, the mistry expressed his disdain for the way the wires had been installed in the first place, and said this was the cause of the trouble.

For his part, Mr. Mehta blamed the wires that had been installed in my apartment when it had been renovated a few years earlier. "They must be Indian," he said, meaning of Indian manufacture, and that explained their failure. To put Mr. Mehta's point differently, "Indian electricals" were notoriously shoddy and represented the weak link in an assemblage composed of private discoms, local workmen, and uneven power supply. But to my eyes, other factors were also at play. When I asked him about the fact that much of the wiring had circumvented the breakers in the apartment, he disparaged the mistry's ability to discern whether or not the wiring was routed through the circuit breakers. Confronted with the wires themselves, so badly damaged and partly burned and melted, Mr. Mehta drew just the one conclusion about their national origin.

After this conversation, Mr. Mehta took me on a tour of his own house and of the electrical goods installed there, which had been unaffected by the power surge. The house had been built by his father in the early 1950s, in the first flush of postcolonial house building as much as nation building. It was a large, sprawling place, with several closed-off additional apartments that belonged to Mr. Mehta's siblings, who now lived overseas. All the electrical fixtures were the same age as the house, and yet, Mr. Mehta told me, he had never had to have any work done on the electricals in the main house. They were all "solid state," and, more important, were imported goods from Germany—they would last for generations. After conducting me to a surprisingly clean cabinet on a lower level of the house, Mr. Mehta pointed with pride to the logo of a German company on the giant, antiquated, industrial-strength switches and breakers that hunkered there. German goods, he said, were quality; the Indian wires that had been used in my apartment, on the other hand, well, they obviously couldn't stand up to Indian conditions.

Enclaves of Contentment

During the period after the dramatic power outage in our neighborhood, Mr. Mehta phoned the main office of BSES Rajdhani, our local discom, for information and updates daily. He also regularly checked in with the teams at work around his property to keep abreast of repair work in the neighborhood. He complained to me about the unresponsive distribution company and the poor coordination of the works that were leaving our streets in a constant state of disruption—always, it seemed, being dug up for one reason or another. Indeed, although I stayed in the neighborhood for several months after the outage, the streets and alleys that surrounded Mr. Mehta's house never regained their paving but remained dug up and disturbed even though there was little evidence of active ongoing work.

In my discussions with Mr. Mehta, he made clear that from his point of view, citywide electrical conditions were palpably deteriorating under privatization. Our neighborhood had been until recently, he said, a "model colony"—a politically privileged area where the state-provided infrastructures worked well and consistently, because of close state supervision and investment of political capital. When Mr. Mehta's father had built the house, the Nehruvian state had provided an implicit political guarantee of efficient operation of modern "amenities" in such neighborhoods, in exchange for the homeowner's own installation of imported electricals. In the acknowledged absence, that is, of the technical capacity to provide universal services, citizens collaborated with the state to construct a local modern plenitude of electrical power through political connections and influence. As Sudipta Kaviraj puts this point, more theoretically, the myriad points of political access and institutions for redistribution that the postindependence Indian state created also carved out, for the middle class that was able to access state benefits and institutions and combine them with their own symbolic and real capital, "a substantial enclave of contentment with the state's performance."[1] It is just such an enclave that was at stake for Mr. Mehta, as it came under threat from neoliberal reform and privatization.

Prior to privatization, that is, Delhi's uneven distribution of power had followed the contours of political privilege, and effective electrical connections had been dependent not only on technical supplies but also on *pahunch* or "reach"—political influence with the local power board engineer, or someone higher up. This contrasted tangibly with privatization, for Mr. Mehta and his class, in that there had been a single source of authority over the grid (one knew whom to talk to), and political favor and not only economic wealth was the basis for access. This created status groups—in model colonies—whose power, authority, and access to electricity were all secured in ways partially separate from their command over economic resources, and whose bureaucratic privilege was symbolically visible in the devices that such privilege maintained. Mr. Mehta's colony had been well connected in all senses, and hence actually received the steady service promised by the universal service ideologies of statist developmentalism.

Under new governmental ideologies of efficiency, privatization had introduced the electricity distribution companies as a new mediator between the state and the citizen, changing the terms of participation most radically for residents like Mr. Mehta (and by association, me)—those who had once benefited from their special "reach." In the privatized economy of power, such reach, and the social proximity to the incumbents of governmental power it was often based upon, no longer mattered. Now, what had been material symbols of status and privilege—including German electricals and responsive local power officers—were superseded by new, more discontinuous patterns of private investment, both by the "discoms" and by individual RWAs. This produced a different, if still partial, distribution of electrical plenitude—one that was, moreover, harder to condense into household installations, and interacted more erratically with them. Worst of all, the lesson of the power surge seemed to be that the "Indian electricals" Mr. Mehta had installed in his rental unit (my apartment) *needed* the kind of high-intensity political investments that privatization was meant to wipe away, in order to function at all.

Mr. Mehta grudgingly adapted to the new conditions, investing his own resources in the repairs but with a regretful recognition of the transformed political economy in which he did so. He pressed the mistry and his team hard, and soon the power in my apartment was back online (although the image on the television never regained its former colors). Once the work was done, Mr. Mehta took me on another tour, this time emphasizing the cost he had gone to in having new, sturdier wires installed, even digging up his own back garden to have a new feeder line laid. However, his pride had dimmed, and the actual appearance of the electricals within my apartment left something to be desired. No longer displaying the care with which he had maintained, and had unveiled to me, the electrical goods that had withstood the outage in the main house—such as the cupboard where his German circuits were mounted, clean and dusted in their antique glory—the tour of the latest improvements was perfunctory, and the "finish" of the newly installed electricals in my apartment was rough. Wires poked through from gaps in the walls where switches had once been mounted, and new plastic conduits were loosely attached at odd angles along the walls. Meanwhile, in the small yard between the house and the apartment, the paving stones that had been broken to lay the new feeder line were loosely fit back together and shifted uncomfortably underfoot. For the duration of my stay, this broken paving remained a scar that divided as much as it joined the two parts of the compound. Now, under transformed political conditions, Mr. Mehta had to invest in electrical goods that were disconnected from any collective purpose beyond the confines of the compound, and they accordingly received less ceremonial attention.[2]

Mr. Mehta certainly did not install German goods in my apartment—that era had passed. He no longer could model, inside his house, an imported modernity as a forerunner of the universality of power and connection that was absent (if still promised) outside it. He could no longer rely on political power or merely local influence within the institutions of state provision to mesh his modern material dreams with those of an embracing, ongoing Indian progress—in the ways that Asaf Ali, for instance, had hoped Indian nationalism might achieve. Rather, within the walls of his own compound, Mr. Mehta had to create a totalized replacement for the power relations of urban modernity with piecemeal materials. As he went about this patchwork labor, Mr. Mehta spoke to me of his own desires and discontents, not addressing himself to the public sphere of formal political ire (which I discussed in chapter 5). But he defined the problems he confronted as having to do with bad materials—Indian electricals—as well as deteriorating exterior conditions. His appeal to a past and irretrievable pattern of state installations and private importations provided a suggestive counterpoint to the complaints about unresponsive authorities and corrupt meters that characterized the public discourse.

For one thing, this new political context meant that Mr. Mehta had to embrace the newly privative form of property in electricity—not only to spend his own money to upgrade his goods but also to insulate himself against external conditions materially rather than through political influence. The "unbundling" of responsibility for local electricity infrastructures encouraged his own local investment of time and energy, while ensuring that his control of local conditions was just that, purely local and diminished in meaning. Yet this new boundedness of political agency meant that when people like Mr. Mehta sought to address wider issues, it was precisely to the neighborhood and to other residents that they turned for solidarity. The neighborhood became a place from which they could effectively act, if no longer as citizens, then at least to create a new political subjectivity based on their possession of property as homeowners and residents. Far from depoliticizing electricity, privatization meant that a property relation became the basis of new political claims. Perhaps Mr. Mehta's grudging investment in the repair of the Indian electricals in his rental apartment was not so disconnected from a collective purpose after all—the rough-and-ready works he had done, in their way, symbolized his limited reach and unwilling participation in a privatized form of governance, and as such provided the affective energy for a new claim of collective rights, based in injured status.

Decaying Status, Collective Energies

One might well ask what pertinence to politics and public life all this talk of German electricals, electronic meters, private investments, and disillusioned practicality could possibly have. How can broken electrical goods, fused wires, and consumer relations with a privatized discom still connect up, even interpretively, with the institutions and practices that made up the "public" grid and shaped its materiality? What relevance could this all possibly have for the longer-term study of political ritual and statecraft in India's modernity? On the basis of this encounter with a power failure and the recovery from it, I venture that the answer lies in the wounded sense of personal status that Mr. Mehta displayed and enacted, as he complained about the downgrading of his neighborhood and its connections to power. That is, the reality of state power here can be sought in a domain of imagination and intellection, where times past and future are commingled with the material realities of the present, and judgments and comparisons—politicized and not—are mobilized to make sense of ongoing transformations. In such moments—interpretive moments—of meaningful action in relation to others and to power, the state is tangibly symbolized in certain kinds of material connection while remaining, immaterially, a promise of the goods—the benefits but also the prerogatives and rights—that will be conveyed across that connection.

Mr. Mehta was still connected to the grid even after privatization—indeed, the works that caused the power outage were meant to improve the quality and regularity of his electric power. And that material connection still mattered to him, even if what it had once meant could only be recalled elegiacally and if the collective promises of the past were now denied by the separating, dividing, privatizing practices of the present. By being translated into something having to do with status and participation, at least within the imagination, the state's (literal, electric) power is lifted above its merely material reality (with its infrastructural supports), to signify high hopes and great aspirations. Moreover, this work of translation, with its shuttling between material reality and moral aspiration, is integral to the legitimation of modern, democratic state power.

As I mentioned in the introduction, the effort in each of my cases has been to describe appearances of technological devices within ritual performances and ceremonies where politicized understandings of status and collectivity were at stake, to see how technological and ritual practices reciprocally inform and shape each other. I drew methodological inspiration from Clifford Geertz's 1980 study of court ritual in nineteenth-century Bali. In that work, Geertz argued that political theory pays far too much attention to material interests and open conflicts over resources and power; instead of this, he advocated attention to the ritual ordering that links these political issues to other centers and sources of meaning, in social understandings and in personal life alike. Geertz writes, "So far as political theory is concerned, it is there, in exposing the symbolic dimensions of state power, that the use of attending to decaying rank, dispersed prerogative, ritualized water control, alien-managed trade, and exemplary cremation lies. Such study restores our sense of the ordering force of display, regard, and drama."[3] On these terms the power that is shaped and conveyed by stately rituals can neither simply be disaggregated into class power, nor analyzed as a displaced effect of more material relations located elsewhere than at the site of ritual performance, nor is this power simply an illusion of verticality with no sociological reality.

To be sure, Mr. Mehta's careful attention to his electrical goods is on first sight simply pragmatic and local and hardly ritualistic or even political—it bespeaks an economic actor's attuned concern for material realities and fears about contingencies. But he builds throughout upon meaningful, emotional investments and transcendent, time-binding expectations, as we can see as we walk with him to inspect his (inherited) German electricals. And more, as he grappled with the local distribution company, the recalcitrant workmen, and complained about the absence of the state, his own devices and installations—his sold-state electricals, his burnt wires—provoked memory, imagination, and politicized comparison with past and present connections, ultimately serving as vehicles for a new negotiation of participation in the symbolic order of the reformist state (a privative and exclusive form of participation, perhaps, but participation nonetheless). In this case, the unexpected damage to his household goods and the lack of public recourse for redress also revealed (to him and to me) how public power and private resources have each, in Delhi, long depended on the other for their fullest realization, and starkly illuminated the new and one-sided political configuration of privative consumership over and above common citizenship.

Let me stress this last point about the intellectual work that technological devices do, in their aspect as ritual symbols, to shape understandings of state power and help to effect its transformations (or reconcile citizens to new statuses). We have no need to assume that because what Mr. Mehta did here transpired in the domain of private action, it had no political significance for him. As Carol Greenhouse has stipulated (in the context of American political life and its ethnography), "The connection between state and citizen is not one of a whole and its parts, but signs, their referents, their necessary fictions, and their ambiguous afterlives."[4] This understanding of the symbolic means by which statehood and citizenship are mutually constituted and transformed brings the work of interpretation to the fore in any account of state power and its relational reality. Through interpretation of ambiguous events and attempts at understanding the distant or obscured causes of proximate effects—as when Mr. Mehta insisted on certain reasons for the failure of his Indian electricals—connections are made between individuals, and the relevance of political structures to their

Figure 7. Contrasts: The last colonial bungalow in Connaught Place under demolition in 2006 (author photo).

lives is realized in action. As we have seen, such signs and fictions (and electronic meters) are routinely enlivened as more than just part of a *local* context, but as integral parts of a *world* in which legal or political power is made into something "workable" in relation to a given project. Greenhouse argues on this basis that our sense of ethnographic context must be redefined to include the reach and scope of people's "embrace" of the state beyond its purely local and its political-formal dimensions, and must include the relatively routine charismatic performances and attempts to know and grasp the present, by which citizenship becomes an active effort to expand relations with other people and to widen the scope of community.[5]

Likewise, Mr. Mehta's response to a local outage reveals his own work of interpretation beyond the immediate givens of the situation, beyond his subjective investments and worry about the value of his own property. Even as he grumbles about the slapdash works in his neighborhood, his intention reaches outward to other people, and upward, invoking the state, its past and promises, and a sense of common, present need. This implicit appeal to political power and citizenship is specific to Mr. Mehta, in his time, generation, and class. On those terms we can sociologically "make sense of" (that is, interpret away) his actions. But, discerned ethnographically, as they reach beyond the confines of sociological context to invoke the past, to interrogate the present, and to claim political belonging, his private worries have public relevance. There is political energy still present within his personal problems with power. That is, such private interactions with electric power inevitably include more than just the "end user" or the "consumer"—they imply a collective subject, insofar as they always extend beyond the boundaries of one subject to encompass the relations and moral qualities that integrate him or her into a wider "grid." In sum, although all the politics of middle-class, post-privatization Delhi, including bhagidari and RWA activism, seem at first to be "merely" a politics of consumption and to concern private actions and property, they are shaped by a history of practices and meanings that guide individuals toward participation in more extensive, more than material, networks.

Mr. Mehta's struggles to grasp his local power conditions indicate that there is still untapped power in the symbols and the affective currents—and the politicized comparisons—that are at play in this history of Indian electricals. Although neoliberalism works to obscure any public power to effect connections and distribute goods, immaterial and material connections remain potent within the urban grid and might yet still be mobilized for wider, more embracing collective purposes beyond the attenuated life of private property and of capital. That will require, however, taking seriously not only the *material* linkages and interdependencies (and disconnections) inherent in modern technological infrastructures but also how these infrastructures have been and are ongoingly used to devise and figure broader collective aspirations. And to be sure, it will also demand a critical intelligence, one that can dismantle the besetting illusions of power and of progress that make it so difficult to tap into the collective energy inherent in technological relations.

CONCLUSION

The Art of a Free Society

The art of a free society consists first in the maintenance of the symbolic
code; and second in fearlessness of revision. . . . Those societies which
cannot combine reverence to their symbols with freedom of revision, must
ultimately decay either from anarchy, or from the slow atrophy of a life
stifled by useless shadows.

—A. N. WHITEHEAD, *Symbolism* (1927)

Morality must leave itself open to repudiation; it provides one possibility
of settling conflict which allows the continuance of personal relationships
against the hard and apparently inevitable fact of misunderstanding, . . . a
way of mending relationships and maintaining the self in opposition to itself
or others. . . . We do not have to agree with each other to live in the same
moral world, but we do have to know and respect one another's differences.

—STANLEY CAVELL, *Claim of Reason* (1979)

This account of electrification and its modern meanings in Delhi began
with Asaf Ali's recollections of being impressed by the "blaze of lights in
the evening . . . uniformity, mechanism, and bustle" on his stopover in
Marseille in 1909. As we saw, Asaf Ali's recollections contain, at once, a
desirous description of technological modernity and a moral apperception
of cultural difference. Such contrastive images of illumination and urban
mores continue to be produced and consumed in contemporary Delhi,
where writers and artists and politicians still construct moral apperceptions

of difference, division, and desire in terms of their public utilities (although nowadays there is no longer any need for people from Delhi to undertake international travel to encounter large-scale urban technologies).

The English-language magazines and newspapers that catered to the affluent middle classes of Delhi in the first flush of twenty-first-century millennial prosperity worked hard to present an image of uniform illumination and urban bustle, of a modern "world city" where urban progress and broad participation in its benefits went hand in hand. A photo layout published under the punning title "Night in Shining Armour" in the October 2005 issue of a glossy Delhi lifestyle magazine, *City Limits*, presented dazzling lights and extensive electrification as facts of everyday experience. The central photograph was a long exposure of the traffic wending its way southward from Connaught Place, or "CP"—the commercial center of New Delhi. The image, taken from one of the tall towers that have risen above CP's colonial arcades, blurred the taillights of the traffic to leave a red-orange fluorescent glow across the center of the photo and to give an impression of traffic moving at speed. The accompanying text hailed the "collective luminance of more than three lakh streetlights, the halogen glare of millions of automobiles [. . .], the metered wattage of this populous city . . ." and breathlessly ended in an ellipsis. After this brief pause, the enthusiasm returned: "Candle power? Even the paan-wallah on a bicycle outside a restaurant and the ice-cream seller at India Gate are touched by the glow on glass and high walls."

This text, complete with its hyperbole and its ellipsis, is especially characteristic of its time and place. The invocation of electrical universality or collective luminance with which it begins is belied by the divisions represented by the "glass and high walls" with which it ends. Meanwhile, these glittering surfaces achieve the remarkable trick of simultaneously excluding the petty urban vendors who populate the margins of this city while also bathing them in the reflections cast by this "shining armour."

Compare this celebration of consumer-age technological urbanity with *Strikes at Time*, a 2011 video installation by the Raqs Media Collective, a Delhi-based group of artists. In this and related works, the artists of Raqs provide a distinctively different visual itinerary through Delhi's global-city modernity. In the main video of this dual-screen work, the camera moves through Delhi's fragmented landscape at night following an eerily luminous vehicle that traverses the wasteland beneath elevated highways and

Figure 8. Image from *How to Get from Here to There* (2011) by the Raqs Media Collective, a work from the same series as *Strikes at Time* (2011). Image Courtesy of Raqs and Frith Street Gallery, London. All Rights Reserved.

circles the billboarded sites of new housing developments. Acid-yellow lights glow above on the highways, while the vehicle itself is outlined (by the artists) by blue-white tube lights; inside it we see people costumed as gods and goddesses riding along as it moves through the city's neglected spaces. Interspersed with these scenes, we witness a laborer writing under the borrowed illumination of work lights or at home in the dim light of a single electric bulb. In the secondary video, his words—composed of reflective meditations, humble daily accounts of inadequate pay and mounting costs, and insights into social inequality—unfurl in elegant Devanagari script.

The slow, balletic movements of the illuminated van across the dismal, dark terrain bring new brilliance to places put in shadow by the "official" lights of the city. The oneiric effect of these images—augmented by the sight of blue gods in work gear traveling in the vehicle—is counterpointed by the mundane reality of the hot work lights and dim domestic glow under which the worker variously writes. The intensity and variegation of the illuminations used in this video point up the weird reality of Delhi's intermittent infrastructures, how they distribute their benefits unevenly.

This video work also seeks to indicate how these given conditions might be reworked in new and transformative ways.

Visually, this piece uses the yellow burn of incandescent illumination as a sensory analogue to the harsh conditions of everyday life for laborers, while the cleaner, bluer light of the dreamlike van literally illuminates the dark waste and destruction wrought by the building of luxury-class modernity in the city. These artful images refract real experience. Light, like land and access to infrastructure, was indeed unequally distributed in Delhi as I experienced it. Private developments set themselves apart from the urban fabric, often illegally tapping into shared resources or expropriating public space to do so. Meanwhile, the informal colonies of the poor found themselves subject to the new judicial power of the discoms under logics of "cost recovery," while multiple privatizations and "pirate modernities" sequestered and enclaved the benefits of electricity, marking the decline of any hope for the technological universality once dreamed of by postcolonial elites. However, in this artwork's vision of the night, with its artificial illuminations, dreams and hopes are still fostered across the manifest divisions—in terms of technological plenitude and possibilities of real participation—that shape residents' and denizens' experiences.[1] Most importantly, the glowing van—a ritualized illumination of dark places—is mobile, and casts its brilliance even where material infrastructure is absent.

Commentaries on the Contemporary

As commentaries on the contemporary, these two images of Delhi's dazzling brilliance differ substantively as well as aesthetically. Visually, the Raqs video lurks at the edges of Delhi's urban landscape and provides dreamlike images of the densities and the lacunae produced by new distributions of power and prosperity. These images can surprise and unsettle in their ambiguity. The text and image interplay in *City Limits*, by contrast, hovers over the center of New Delhi and promises mastery of the "metered wattage" of the city. It invites critique by all too plainly revealing (and reveling in) the bad faith of a new state dispensation of consumer-citizenship, which proliferates exclusions while promising a universe of commodities in lieu of participation.

Yet if the latter image seems to offer a kind of universality and inclusion that is, at the same time, technologically and materially illusory ("the glow on glass and high walls"), the comparison with the Raqs Collective's piece might push our thought further and in a different critical direction. To be sure, it is intellectually satisfying to strip away pleasing, aesthetic illusions to leave behind only the naked reality of division and separation. As the anthropologist A. M. Hocart wrote long ago, when we moderns gaze upon modern urban scenes of technology and magical, glittering display, the voice of the "rationalizing historian" chimes in to ask, "Will you maintain that these [illuminated] heavens are anything but a play of fancy, but pure art?" However, it is important that Hocart rejected this disillusioned vision as unhelpful for understanding the practical and ultimately political work that is undertaken in ritual and through aesthetic elaboration of the mundane facts of our everyday existence. He argued that an electrically illuminated heavens or a ritually adorned utilitarian device do not only fool the eye or entice the imagination into mirrored realms of illusion; they figure connections beyond the presently tangible and can be part of a "scheme to secure life."[2]

Strikes at Time, set in counterpoint to the bad faith of the glossy lifestyle magazine's image of Delhi's electrical modernity, does not only reveal abjection, does not only spur us to materialist critique. It represents, in a different register than that offered by the mass-circulation media image, a play of fancy that opens new insights into the conditions of the contemporary and a constructive deployment of frankly illusionary images. These artists literally illuminated the spaces otherwise only dimly visible under the reflected glow of the city's official lights. Their illuminated van doubles aesthetically for the quite concrete and material efforts to appropriate light and power that on one level structure the reality of Delhi's grid and its citizens' relations to power. But by offering an aesthetic reduplication of material processes of both exclusion and appropriation, it does more than point up unequal material conditions; it also offers a reminder that a great deal of energy remains in the purely imaginational work that can be spurred by borrowed, reflected, or intermittent illumination.

The piece by Raqs interrogates, through the magical appearance of *light out of place*, how a modern, democratic state's immaterial promises of inclusion and participation are at one and the same time belied by and refracted within the diurnal routines of city life. What it shows us, that is, is an irresolvable and aporetic encounter between the mundane, the "everyday" level of action (with the worker's diary) and the ritual promise and power of hallucinatory images cast across that real experience. This work, with its complex interweaving of reflection and reality, may help us think beyond a simple critique of the *inevitable* gap between, on the one hand, political promises of progress or liberal ideologies of belonging and, on the other, hard, material realities; to consider instead how this gap or hiatus is experienced from within, its limits tested, and its form continually re-created and potentially transformed.[3]

The cases of electrification and politics in this book offer, on these terms, an insight into how, pragmatically, moral thought and ritual performance engage with and grapple with technological devices (and the infrastructures to which they connect) and offer other ways of imagining community and participation when infrastructures and devices merely beset and exclude (as they so often do). This work is always ongoing and involves symbolic elaborations of categories and codes of understanding (through significant devices and political statuses) as well as moral critiques of the material conditions of the present.

Any cultural work with power and light, meanwhile, is incessantly subject to revision and repudiation, as any meaningful politics worth the name must be (see the twin epigraphs to this conclusion). In sum, neither a legal, formal order of rights and duties that governs state and citizen, nor the material conditions of everyday life that equally impinge upon action and thought, can be understood on their own, without the "irreducible relation of culture" that links them together in a whole, lived, social reality.[4]

Moralizing Politics, Again

In this book, moral moments in the history of electrification in Delhi have been found when and where electrification provided the occasion for actors to achieve some "grasp of the wider predicament" in which they found themselves, politically, and to ask collectively—in relation to the government or to other citizens—"how we continuingly stand or have stood to others and to power."[5] While the philosopher Charles Taylor articulated this question and its urgency (how *do* we stand in relation to others and to power?) in the course of a discussion of secularism and democratic citizenship, some such question is not the exclusive preserve of the avowed modernists in this history; moreover, even disenchanted technocrats need this kind of idiom of collectivity and commonality in order to express their highest concerns with order and justice. This question about standing, about political stature and one's connections to others—phrased technologically—echoes in the political thought of all the major figures discussed here, in different idioms and with distinct intentions, giving political substance and moral meaning to mundane matters of the scope, timing, and means of electrification.

Each period of politics surveyed in this book, in fact, is predicated on a specific (and cross-comparable) harnessing of glory to order, linking the power of ethical mastery over the political world to the material distribution of roles and responsibilities, so that both together may foster collective life. Even the imperialist viceroy George Nathaniel Curzon pursued a politics that was infused with moral force and meaning, even if his was a wholly narcissistic and self-aggrandizing ritual repertoire (see the picture in chapter 1 of him atop an elephant). Curzon was not only the sly impresario of a colonial sublime, we might say, but also an intent ritual specialist seeking to create intangible bonds through material boons, when he staged his giant durbar and sought to produce loyalty by the ceremonial announcement of a tax remission. He too thought he was building something more than a "play of fancy" with his temporary durbar city, in all its technological splendor. Now we might be in a position to aver, like Hocart, that a play of fancy is not such a small achievement, after all, and its real political implications not so easy to dismiss.

The necessary entanglement of the moral and the material in modern political action—and more particularly the way this founds and sustains large-scale political orders—is perhaps best exemplified by the contrasting but complementary figures of Gandhi and Nehru. For all their close intellectual partnership and emotional bond forged in the decades of the independence struggle, these two political figures are conventionally treated as exemplars of opposing tendencies in modern Indian political practice and thought. As we saw in chapter 2, however, the anthropologist Milton Singer noted, during fieldwork in the 1950s, how often the "religious" and ascetic register of Gandhian politics was integrated into the worldly projects of material transformation pursued by Nehruvian elites. For Singer, the distinction between Gandhians and Nehruvians was hard to draw, and certainly could not be shoehorned into a politicized opposition between tradition-minded conservatives and modern-minded men and women concerned with time and the city. Rather than deal with such static alternatives, Singer stressed how Gandhian symbols expressed, and organized, collective orientations to equality and freedom that were equally central to Nehru's vision of industrialized modernity. Let me quote his observation in full, once again. "Gandhi's campaign for weaving hand-spun *khadi*, for example, did not succeed in replacing factory looms with the ancient spinning wheel. Yet it does seem to have succeeded in dramatizing concern for cottage industries, the dignity of hand labor, and village underemployment, while providing a symbol for a self-respecting cultural identity in a successful mass political movement. From this point of view, one might argue that the *charkha* (spinning wheel) really articulated in an archaic idiom the voice of Congress and of the sewing machine."[6] Gandhi's spinning wheel or charkha was, Singer's observation reminds us, just as much a technological tool of material participation and a collective political symbol as it was the means of his humble, ethical, ascetic practice. Meanwhile, Nehru's industrial-scale "new temples"—dams, power plants, and other massive signs of energetic capacity—were in turn ritualized as symbols of India's collective moral purpose and destiny, and themselves provoked ethical and even (as we have seen in chapter 3) poetic interpretations.

To rephrase Singer's insight in the terms used by Giorgio Agamben to describe the opposed poles of political theology and governmentality, and their joint contribution to the machinery of modern statecraft, distinct visions of otherworldly glory are knit together with institutions of worldly order and distributions of common goods to produce the architecture of the modern state. Gandhi's utopian vision of cottage production was just as much an image of political order and economic ordering (a deeply conservative and moralizing one) as it was a religious one, and it was possible to bring it into play as a ritual orientation—an ascetic, self-sacrificing one—throughout the project of postindependence state formation.

The third case of privatization and neighborhood politics in Delhi is distinct in two important regards from the earlier ones, since it is not nearly so focused on high state politics, charismatic leaders, or ritual events, and its rituals and devices seem to be about division and separation—walls and meters—more than moral orientation and ethical connection. At the time of my fieldwork, the politics of privatization offered no outstanding charismatic figure who seemed to sum up the contemporary logic of distribution or the symbolic economy of power, or to personally articulate new configurations of glory and order (through repudiation and revision of what had come before) under conditions of privatization. The later emergence of the anticorruption protests of 2011, and Arvind Kejriwal's subsequent rise to political prominence with his language of *satyagraha* and practice of ritualized protest (or *dharna*), and his mobilization of a wider swath of Delhi's population around issues of fairness and inclusion in the city's amenities, indicate that the associational politics of 2005 and 2006 did indeed fuel a new wave of moralized politics. Indeed, the anticorruption politics at the outset shared personnel and procedures with the post-privatization politics that I described in chapter 5. However, we do not need to look so far, or find exceptional individual politicians, to make comparisons with the earlier eras of electricity politics in Delhi. In 2005, the RWAs themselves constituted sites of a new collective charisma—albeit an impersonal one—transforming the terms of belonging in Delhi through their vigorous action to create new walls, gates, and associational boundaries.[7] A further impetus to this moralized politics came with the advent of the new metering technologies and the newly divided interests in the grid that they represented. The market-based divisions and disciplines that were imposed on the old political connections of the public utility spurred countervailing, albeit newly bounded, forms of collective organization.

Through their gating and segregation of neighborhoods, the RWAs also reproduced nearly colonial techniques of separation and indifference, and the inequalities and divisions of the contemporary city appear to reproduce an imperial logic of status differentiations. But this cannot be read as a simple "return" to antique logics: the fact that any collective force was available to be mobilized—and that it was precisely the entanglement of homes and neighborhoods in a shared grid that contradicted the individualizing logics of privatizations—underscores just how important all the past material and moral settlements and distributions of power have been in shaping the current politics of electrification in Delhi. When new electricity laws took force in Delhi, instituting new divisions, multiple affects and conceptions of citizenship were already bound up in the electrical infrastructure, and a new sense of collectivity emerged in the attempt to enforce reformed relations to the grid through new electronic meters. Thus, the politics of the RWAs were not just about status-rich distinctions, and not just the expression of neoliberal imperatives of reform, but rather an active cultural politics that grappled with both the past and the present, recasting connectivity once again, despite privatization, as an act of public participation (now, however, limited to the neighborhood level, to "model colonies"). The reform of electric power in Delhi represents another turn in a comprehensive affective shaping of technological infrastructures of provision, which makes them reworkable emblems of collectivity and belonging; this emblematic function is one key sense in which electricity infrastructures and devices are "moral technologies," and one that is evident across all the cases here. Still, the imaginative scale at which the infrastructures of Mr. Mehta's Delhi operate, the moral scope of the work done with them, is definitively narrower than those that inspired imagination and action in Srinivas's Rampura, while returning to something more like the temporary installation and sumptuary elaboration of infrastructures that defined imperial electrification.

Electrical Community

What of the future? Will Delhi's state and citizens secure 24/7 power? For some citizens, this is already a reality—buffered from the grid by their own private devices or able to enjoy the new, middle-class amenities of Delhi's malls and luxury developments, some of which have dedicated power generation. More than once I went to the movies in Delhi (in the luxurious Saket shopping center, even), and the power went out partway through the film. That no longer happens in the new malls and most affluent colonies.

This already indicates that 24/7 power, when it is a reality, will still not be universally available on the terms or through the technological means by which it was first imagined. A freely flowing, inclusive electrical modernity was projected by the Congress planning commission and by postindependence villagers as a product of central stations, national grids, good citizenship, and state coordination. Such comprehensive coordinations of power and belonging are neither politically, economically, nor ecologically desirable now. So what does contemporary Delhi offer as a pathway forward? Anticorruption protests, walls and gates, and private enclaves of plenty? Meanwhile, a nationwide politics of communal division—between Hindu and Muslim—and other identitarian politics continue to shape the public sphere, as do multiple other distinctions and divisions (not least between affluence and poverty). I want to close with a final reflection on the notion of an "urban community"—as an anthropological problem, not an ethnographic datum—in order to make a few more general theoretical claims: first, about the ritual processes and legal procedures through which we have seen disparately scaled collectives being formed across this history of electrification in Delhi; and second, how this history illuminates the terms on which an embracing, inclusive urban community may yet be constituted and participate, collectively, in making Delhi's future. To anticipate my conclusion, acts and practices of division, if they are ritual and not merely material, may be more important for the formation of political community and even universal participation than at first seems likely.

Ethnographically, "community" was not a key term in any of the contexts where I conducted research. For all their charismatic actions of gating, walling, and segregating themselves from the city, and mobilizing on the basis of residence and shared interests, RWAs and the advocates of privatization in Delhi made scant use of the word—though frequently invoking related notions of collective rights and moral order. There are good reasons for this absence. In Indian English at least, "community" has a complex history. As Indian religious, caste, and tribal "communities" were identified by colonial knowledge, enrolled in colonialism's efforts to forge governable units out of India's diversities, and later further reified by the nation-state's categorizations and discriminations, supposedly primordial "communities" were typed as at once the residue of India's premodern history and the basis of its permanent political difference.[8] "Communal," even today, is all but synonymous with religious antagonisms, even violence. Governed by colonial genealogies and by the master image of partition and the communal electoral and social politics that produced it, then, community is a deeply loaded word in postcolonial India, and some readers may understandably have been surprised at my occasional uses of it throughout this book as an analytic term to describe the partial, differentiated solidarities that have been imagined across India's modern political history.

Delhi's political status and stature as capital city, as we have seen, were likewise overdetermined by both colonial and indigenous assessments of the city's communal identity. The city symbolized Mughal sovereignty in the eyes of the colonial ritual specialists who first installed electric power there, while it was as a *Muslim* congressman that Asaf Ali found a political constituency in Delhi and acted within the larger nationalist movement. The post-partition politics of home rule sought to remake the city as the natural home of distinct language groups and ethnic communities, as we heard in the debates from the Constituent Assembly. In both these cases, again, claims of community appear to relate back to some natural, primordial organization of society, principles of group identity and social differentiation that can be mobilized for the purposes of government.

The notion that community properly relates to and involves primordial principles of group identity, based on exclusive differences, has been challenged by both postcolonial and postmodern thought. Dipesh Chakrabarty, thinking postcolonially, has sought to reclaim the word itself from communalist rhetoric, and to remind us of the richness and embracing religious sensibility of a premodern world populated by many and diverse communities.[9] He has said, "I see community as an always-already fragmented phenomenon"—refusing any recuperation of community in a logic of identitarian sameness or immutable difference, or any notion of a collective subject bearing historical continuity from past to present.[10] With this, Chakrabarty echoes, again postcolonially, the deconstructive rethinkings of the term by European postmodern thinkers like Maurice Blanchot, Jean-Luc Nancy, and Roberto Esposito, who have emphasized in a different context the historical and philosophical fact of any community's dependence not only on a claim of identity but also, even more fundamentally, on notions of divided ownership and shared possession. Esposito and Nancy both interrogate the etymological origins of the very word, revealing the incessant division implicated by its historical meanings of sharing out or dividing up, owning *with others*. Community, these postcolonial and postmodern writers have all insisted, is "no-thing," not the positive object of any sociology, and as such should always present a challenge, an aporia, to any political thought that would seek to found or institute an undivided and unitary *res publica* (a public thing).[11]

These latter problematizations of community (which are much broader than I can address here) have stressed, as an alternative to positive visions of community made through assertions of unity and primordial identity, the productivity of thinking anthropologically about community through the more active, more fundamental concept of *division* and in terms of a group's nonidentity and non-positivity. Meanwhile, the related postcolonial literary and historical efforts to define more pragmatically the possibilities of an "anti-communitarian communitarianism," a radical and yet common "co-belonging of non-identical singularities," have engaged concepts like "friendship" (in Leela Gandhi's critical work on anticolonial thought) and Bengali practices like *adda*—an intellectual, urbane sociability excavated historically by Chakrabarty—as the basis for a different, nonidentitarian and non-primordial, sensibility of collectivity and of co-implication in action.[12] These latter explorations of affective and vernacular practices, with their alternative and anti-communitarian ways of envisaging sociality and durable moral bonds, obviously provide no basis for any "anthropological" claim to draw lessons from a distinctively "Indian" historicity or experience, because of the fatal implication of a historically continuous communal identity that that would bring. But they do seem to offer some basis for rethinking how urbane, technological practices of connection (and, ineluctably, also therefore division) and primordialist and communitarian *claims* of historical unity and common progress have interacted over time. Leela Gandhi's and Dipesh Chakrabarty's explorations of the production (and, at times, eclipse) of cosmopolitan, modern communities marked by their essential divisions highlight and try to reclaim—as I have tried to do, also—undirected and open-ended ritual reworkings of division and distinction as the paradoxical basis for a more humane and more democratic collective historicity. Finally, acts of division inaugurate a sociologically impossible but still intellectually (and politically) necessary whole and allow us to discern how discrete, technological actions impinge on others and organize relations, distributing power and regulating social action.

Envoi: An Ancient City

Delhi's antiquity, its religious and political history, its subjection to colonial divisions and discriminations, its both grand and violent remaking over the first half of the twentieth century, and its reconfiguration under novel political logics of decentralization and privatization in the twenty-first, all finally converge upon one paradox: division creates the possibility for any politics of community, of sharing out and participating in the goods of a life in common. In New Delhi's techno-political history, meanwhile, ritual performances have both produced its centrality—as capital city—and fomented, internally, divisions and distinctions. These performances have been, in each case, also the site for the production of meanings and practices that rework, reinstrument, and renew urban technological modernity.

We may find deeper intellectual (and anthropological) foundations for our paradox of ritual division at the heart of "community" in the work of a nineteenth-century classical scholar who taught Émile Durkheim and was influential in anthropological theorizing throughout the twentieth century. In his 1864 masterpiece *The Ancient City*, Numa Denis Fustel de Coulanges (usually referred to in the literature as "Fustel") sought to understand the archaic roots of ancient Mediterranean cities (and hence, for him, the very kernel of all civilization). Fustel asked what could have achieved the seemingly miraculous transition from primitive and disconnected groups of people to the remarkable feat of political order that was the ancient city, with its strong solidary unity and its differentiated rights and obligations. At the origins of urban life, he wrote, something had to be present that was "stronger than material force, more respectable than interest, surer than a philosophical theory, more unchangeable than a convention; something that should dwell equally in all hearts and be all-powerful there."[13] He meant that it would be fruitless to seek the kernel of this ancient political organization in an exchange between pre-existing groups, a contract, a conquest, a mutual and reciprocal respect for property, or indeed in any of the other places where political philosophers had sought the positive beginning of the artificial bonds of political society.

Fustel observed, rather, that at the foundation of cities one finds, again and again, ritual practices of division, separation, and distribution, arts of sharing out the goods of life, to create the urban community of shared interest and belonging. As Fustel wrote, "If a legislator undertook to establish order among . . . men . . . he never failed to commence by dividing them into tribes and fratries, as if this were the only type of society. In each of these organizations, he named an eponymous hero, established sacrifices, and inaugurated traditions." With this work of inauguration, not despite but because of its institution of an artificial order of ritual and mode of association ("as if this were the only type of society"), division became the singular basis for the unity of the newly defined parts. This work achieved a social bond both stronger and wider than any that could have been created by additive or filiative processes alone.

Within the present-day political context of a city like Delhi, where territorial divisions proliferate as the basis for separate and jealous claims on resources and on rights to the city, Fustel's thesis offers a challenging but necessary revision to our understanding of both the material bases and the solidaristic possibilities of urban community. With his thesis about the ritual origins of cities through the inauguration of difference and division, Fustel forestalls any approach that takes these divisions for granted, as primordial, or as the inevitable result of economic or political processes. *The Ancient City*, thus, helps foster a more radical thought about the nature and quality of the modern, infrastructural social tie, which also separates as it unites. We can, guided by this insight, seek and find in ritual inaugurations and meaningful installations the very criteria of division and distinction by which cities are organized politically and materially, and on this basis we may find it possible (if not easy) to seize the opportunity, when it arrives, to repudiate or revise these criteria.

The fundamental practice of social division and distinction undergirds, anthropologically, all the history related here of electrical displays, installations, expansions, and besetting organization of electrical grids through law and politics. This is why thought about electricity and electrification has served here as such a useful vehicle for examining India's political history. Electrical devices symbolize and materially effect, in the thought and practice of myriad actors, both division and connection, both distant and illusionary hope and concrete immediate need; electrical connections and the quality of the energy they supply offer *one* important, material site for the moralization of common needs and the articulation of new collective claims on power (as well as the creation of new boundaries).

Think, one last time, of the compelling images of electrical life in Delhi by the Raqs Collective and in the magazine *City Limits*. Both these works highlight differences and divisions while showing that reflection and borrowing and imaginational appropriation can and do constitute another dimension of Delhi's urban modernity. They both, in different ways, strive to represent dynamic processes of inclusion that cut across the purely mechanical growth of a grid of technological relations. They reveal electrification as political ritual.

NOTES

Abbreviations

AMP Albert A. Mayer Papers, University of Chicago Library
CCO Chief Commissioner's Records, Delhi State Department of Archives
DCO Deputy Commissioner's Records, Delhi State Department of Archives
DSA Delhi State Department of Archives
IOR India Office Records, British Library
NAS National Records of Scotland (formerly, National Archives of Scotland)

Preface

1. As a meaningful politics, this kind of ambient presence and irruptive imposition of infra-structures, affecting both social expectations and personal desires, has been explored by a range of scholars, from literary studies, media studies, and anthropology. Bruce Robbins analyzed the pres-ence of infrastructures in literary modernism in his programmatic essay, "The Smell of Infrastruc-ture," *boundary 2*, no. 34 (2007): 26–33, an inquiry further followed through in the case of Irish literary modernism and global literatures by Michael D. Rubenstein, *Public Works: Infrastructure, Irish Modernism, and the Postcolonial* (South Bend, IN: Notre Dame University Press, 2010). The anthropological literatures, especially on the sensory politics of infrastructures, have been astutely surveyed by Brian Larkin, in "Politics and Poetics of Infrastructure," *Annual Review of Anthro-pology* 42 (2013): 327–43. More recently, Douglas Rogers has eloquently analyzed the production of new senses of rootedness, cultural specificity, and depth in the parallel case of energy—here

oil—politics in post-Soviet Russia, in *The Depths of Russia: Oil, Power, and Culture after Socialism* (Ithaca, NY: Cornell University Press, 2015).

2. This is not the place for an exhaustive literature review, but let me simply offer as an example of this kind of argumentation Timothy Mitchell's brilliant exposé of how the production and management of carbon flows materially shape global politics; he dives deep into actions along the line of production and consumption to account for both state power and resistance at a local, enacted level and writes explicitly that explanations of infrastructural political powers have no need to "detour into questions of a shared culture or collective consciousness." See Mitchell, *Carbon Democracy: Political Power in the Age of Oil* (New York: Verso, 2011), 21. Mitchell, like many scholars pursuing this line of thought, draws insight from Bruno Latour's "actor-network theory." Though I endorse Latour's well-known rejection of purely modernist bifurcations of the world into mutually un-communicating domains of "technology" and "morality," I disagree that anthropological theory is particularly complicit in this. Accordingly, I have not followed his methodological resolution of the two domains into an entangled unity where the technological ultimately takes explanatory priority, as agency (or qualities of action) are expanded to encompass both human and nonhuman actors. See, e.g., Bruno Latour, "Morality and Technology: The End of the Means," *Theory, Culture, and Society* 19 (2002): 247–60. I do not share, either, Latour's stern rejection of both collective consciousness and any social explanation as abstract fictions. See *Reassembling the Social: An Introduction to Actor-Network Theory* (New York: Oxford University Press, 2005). That said, Latour's actor-network theory has been exquisitely deployed to understand the silent technological crafting of distinctions between groups in the context of Palestine, by Ronen Shamir, in his *Current Flow: The Electrification of Palestine* (Stanford, CA: Stanford University Press, 2013).

3. See, for further argument on this point, Leo Coleman, "The Imagining Life: Reflections on Imagination in Political Anthropology," in *Reflections on Imagination: Human Capacity and Ethnographic Method*, ed. Mark Harris and Nigel Rapport (Burlington, VT: Ashgate, 2015), 195–214.

4. Although I was unable to use their work in writing this book, I should note that Penny Harvey and Hannah Knox have highlighted inaugurations—political and ritual events—as important for understanding the meaning of roads as infrastructure in Peru, in *Roads: An Anthropology of Infrastructure and Expertise* (Ithaca, NY: Cornell University Press, 2015).

5. A. M. Hocart, *Kings and Councillors: An Essay in the Comparative Anatomy of Human Society* (Chicago: University of Chicago Press, 1970), 230–32.

6. A. M. Hocart, *Kingship* (Oxford: Oxford University Press, 1927), 41.

7. Isabelle Stengers, *Thinking with Whitehead: A Free and Wild Creation of Concepts*, trans. Michael Chase (Cambridge, MA: Harvard University Press), 136.

8. Ibid., 333.

9. As Marc Abélès has pointed out, Hocart's "line of reasoning seems like a sort of arbitrary presupposition: either the need for [some] kind of organizational form arises and so it actually comes into existence; or this need does not arise, and as a result the organism in question remains in the fictive domain." However, Abélès goes on to say that the association between ritual and (bio) politics that buttresses Hocart's historicist hypothesis does demand further attention. Abélès, *The Politics of Survival*, trans. Julie Kleinman (New York: Polity, 2010), 126–27.

10. This argument is most fully presented in *Kings and Councillors*.

11. Hocart, *Kingship*, 45.

12. João Biehl has used the term "moral technologies" in another context to refer to pharmaceutical regimes and medical institutions that seek to intervene directly upon the psyche, to shape individual behavior and reshape wider, social affective economies; see Biehl, *Vita: Life in a Zone of Social Abandonment* (Berkeley: University of California Press, 2004), 8. I intend "moral technology" in a wider, more Durkheimian sense, emphasizing how technologies reveal and make

intellectually accessible social relationships and in turn are shaped by the evaluation and ongoing cultivation of such relationships. Cf. Leo Coleman, "Infrastructure and Imagination: Meters, Dams, and State Imagination in India and Scotland," *American Ethnologist* 41 (2014): 457–72.

13. Stanley Cavell, *The Claim of Reason* (Oxford: Clarendon, 1979), 268.

Introduction. Electricity Acts

1. *M. Asaf Ali's Memoirs: The Emergence of Modern India*, ed. G. N. S. Raghavan (Delhi: Ajanta, 1994), 67.

2. See Narayani Gupta, *Delhi between Two Empires, 1803–1931: Society, Government, and Urban Growth* (Delhi: Oxford University Press, 1981), 151; Sangat Singh, *Freedom Movement in Delhi, 1858–1919* (New Delhi: Associated Publishing House, 1972), 330.

3. Raghavan, *M. Asaf Ali's Memoirs*, 68.

4. According to the historian Manu Goswami, the dissemination in "the socio-encyclopedic annual *Moral and Material Progress Report* . . . of official statistical and economic figures and representations . . . conjured, in specular fashion, a vast collection of commodities as metonymic of a dynamic and improving colonial economy . . . [and this] logic of visualization worked as a simulacrum of political consensus." See Goswami, *Producing India: From Colonial Economy to National Space* (Chicago: University of Chicago Press, 2004), 80. William Glover calls this effect the "materialist pedagogy" of urban technique, in *Making Lahore Modern: Constructing and Imagining a Colonial City* (Minneapolis: University of Minnesota Press, 2008), vii. For further historical investigation of the colonial ideology of "material and moral progress" see Michael Mann, "Torchbearers on the Path of Progress: Britain's Ideology of 'Moral and Material Progress' in India; An Introductory Essay," in *Colonialism as Civilizing Mission*, ed. Harald Fischer-Tiné and Michael Mann (London: Anthem, 2004), 1–29, and Leslie Sklair, *The Sociology of Progress* (Boston: Routledge & Kegan Paul, 1970), 39–41.

5. The anthropologist Tim Choy argues in a different context that such "politicized acts of comparison" between immediate, present-day experiences and conditions elsewhere or in some imagined past or future can provide the impetus for new and transformative understandings of society and belonging. I have borrowed Choy's phrase freely throughout this argument. Tim Choy, *Ecologies of Comparison: An Ethnography of Endangerment in Hong Kong* (Stanford, CA: Stanford University Press, 2011), 6.

6. Rudolf Mrázek's *Engineers of Happy Land: Technology and Nationalism in a Colony* (Princeton, NJ: Princeton University Press, 2002), a history of technological sensibilities and collective dreams of freedom in Indonesia, provides the fullest explication to date of this imaginary in a related colonial context.

7. Compare Akhil Gupta's classic account of the "circulation" of ideas and representations of the Indian state alongside material projects of development, and how these together shape local understandings of its powers—both as he originally presented the argument in "Blurred Boundaries: The Discourse of Corruption, the Culture of Politics, and the Imagined State," *American Ethnologist* 22 (1995): 375–402, and as it is rehearsed (and slightly updated) in *Red Tape: Bureaucracy, Structural Violence, and Poverty in India* (Durham, NC: Duke University Press, 2012), 75–110.

8. Sunila S. Kale, *Electrifying India: Regional Political Economies of Development* (Stanford, CA: Stanford University Press, 2015); quotes on pages 9 and 31.

9. For recent analyses of the anticorruption politics in Delhi as a ritual, charismatic, and aesthetic politics see Erica Bornstein and Aradhana Sharma, "The Righteous and the Rightful: The Technomoral Politics of NGOs, Social Movements, and the State in India," *American Ethnologist* 43 (2016): 76–90; and Martin Webb, "Short Circuits: The Aesthetics of Protest, Media, and Martyrdom in Indian Anti-Corruption Activism," in *The Political Aesthetics of Global Protest*, ed.

P. Werbner, M. Webb, and K. Spellman-Poots (Edinburgh: University of Edinburgh Press, 2014), 193–221.

10. Aradhana Sharma, "Epic Fasts and Shallow Spectacles: The 'India against Corruption' Movement, Its Critics, and the Re-making of Gandhi," *South Asia* 37 (2014): 365–80. See also Partha Chatterjee's careful analysis of the moral force of Gandhian mobilization, and its ambiguous power within the nationalist movement as a whole, in his *Nationalist Thought and the Colonial World: A Derivative Discourse?* (London: Zed Books, 1986): 113–25.

11. See Navroz Dubash and Sudhir Chella Rajan, "Power Politics: Process of Power Sector Reform in India," *Economic and Political Weekly* 36 (2001): 3367–87 and 3389–90.

12. I should say here that corruption itself appeared in my research more as rumor and supposition than evidenced reality.

13. N. Gupta, *Delhi between Two Empires*, 175.

14. Report of the Delhi Electric Supply Enquiry (Pitkeathly) Committee, vol. 2 (1937), 7. IOR V/26/740/4. See chap. 2 below.

15. This was the methodological presupposition of a World Bank–sponsored survey that was discussed in Delhi during my fieldwork. See V. Santhakumar, *Analysing Opposition to Reforms: The Electricity Sector in Delhi* (New Delhi: Sage, 2008).

16. It is important to note that technological historians have long been deeply attentive to the role played by social and cultural forces in technological change. Thomas P. Hughes's magisterial work *Networks of Power: Electrification in Western Society, 1883–1930* (Baltimore: Johns Hopkins University Press, 1983) explores in depth the social, legal, and institutional forces guiding technological development and sets the standard and the template for much later historical work in this vein (an Indian exemplar of this approach is Srinivasa Rao and John Lourdusamy, "Colonialism and the Development of Electricity: The Case of Madras Presidency, 1900–1947," *Science, Technology & Society* 15 [2010]: 27–54). For Hughes, the historical problem was that the rate and pace of technological change varied between nations, and hence any history of innovation had to attend to social and historical forces at work within a technological system, in the forms of law, regulation, and politics. As Brian Larkin has recently noted, "For Hughes, a holding company or an accounting practice is as much a technical invention as is a dynamo or a telephone," and its development is just as much part of the historical problem as the invention of the lightbulb. Brian Larkin, "The Politics and Poetics of Infrastructure," *Annual Review of Anthropology* 42 (2013): 330. However, this approach does limit the range of questions one can ask to historical ones about timing, pace, and sociological organization of technological installations, and even the much more social history provided by David Arnold, in his *Everyday Technology: Machines and the Making of Indian Modernity* (Chicago: University of Chicago Press, 2013), still puts priority upon when technologies are adapted and how they are used (placing more emphasis on their availability as resources for resistance) rather than the thought they provoke. For a perspicacious review of work on the "technologization of society" in the West, highlighting its inveterately historicist orientation, see Rosalind H. Williams, "Our Technological Age, from Inside Out," *Technology and Culture* 55 (2014): 462–64.

17. Sudipta Kaviraj, "On the Enchantment of the State: Indian Thought on the Role of the State in the Narrative of Modernity," *European Journal of Sociology* 46 (2005): 285.

18. Quoted in Kale, *Electrifying India*, 28.

19. Shiv Visvanathan, "Between Cosmology and System: The Heuristics of a Dissenting Imagination," in *Another World Is Possible: Beyond Northern Epistemologies*, ed. Bonaventura de Sousa Santos (New York: Verso, 2008), 193.

20. Dominic Boyer, "Anthropology Electric," *Cultural Anthropology* 30 (2015): 533.

21. Visvanathan, "Between Cosmology and System," 194.

22. This "other dimension" is perhaps what Boyer refers to when he writes of "the all-too-human histories of power and enablement" that may be opened by an "anthropology electric"

(p. 532). This kind of promise of participation and inclusion has hardly gone unnoticed in anthropological analyses of Indian modernity—and has been identified as among the "enchantments of modernity" by Saurabh Dube in his edited volume of the same title (Chicago: University of Chicago Press, 2009).

23. This is not to say that biopolitics, though often analyzed in its dimension of exclusionary "letting die," does not also "include." As is made clear shortly, we are talking about processes that interact, and are both processes of control in a larger sense.

24. Mary Douglas, *Cultural Bias* (London: Royal Anthropological Institute, 1978), 14. Douglas defined and promoted several versions of "grid-group" analysis over her career; I am drawing here only on that version articulated in *Cultural Bias*. See James V. Spickard, "A Guide to Mary Douglas's Three Version of Grid/Group Theory," *Sociological Analysis* 50 (1989): 151–70.

25. Douglas, *Cultural Bias*, 54.

26. Ibid., 6.

27. The question of whether such ideological ceremonials are in fact meaningful and to whom is a fraught one, of course, which will be addressed in more detail in the relevant chapters. Let me say here that just this question of the relation between governmental efficacy and cultural power has informed all the anthropological debate over colonial rituals, from Bernard Cohn to Douglas Haynes to Andrew Apter (to sketch an intellectual genealogy), each of whom deals with the problem of the imperial ideological content and subsequent transformative embrace of ritual forms by colonized populations in slightly different ways—it is for each a larger part of their historical task to trace these latter, longer-term effects and transformations. I am less focused on these historical-cum-cultural effects of ritual and more on the inverse, but related, question of how distinct ritual formations impinge locally and contemporaneously upon material practices of government. See Bernard Cohn, "Representing Authority in Victorian India," in *The Invention of Tradition*, ed. E. J. Hobsbawm and T. O. Ranger (New York: Cambridge University Press, 1983), 165–209; Douglas Haynes, *Rhetoric and Ritual in Colonial India: The Shaping of a Public Culture in Surat City, 1852–1928* (Berkeley: University of California Press, 1990); and Andrew Apter, *The Pan-African Nation: Oil and the Spectacle of Culture in Nigeria* (Chicago: University of Chicago Press, 2005).

28. Sally Falk Moore and Barbara G. Myerhoff, "Secular Ritual: Forms and Meanings," in *Secular Ritual*, ed. S. F. Moore and B. G. Myerhoff (Assen, Netherlands: Van Gorcum, 1977), 7.

29. Ibid., 19.

30. See the quote from de Certeau used as an epigraph to this introduction, and also Fredric Jameson's defense of the notion of a "singular modernity" as a heuristic in literary studies. Jameson writes, "Modernity always means setting a date and positing a beginning." Fredric Jameson, *A Singular Modernity: Essay on the Ontology of the Present* (New York: Verso, 2002), 29. Michel de Certeau, "History: Science and Fiction," in *Heterologies: The Discourse of the Other*, trans. Brian Massumi (Minneapolis: University of Minnesota Press, 1986), 220–21.

31. *The Education of Henry Adams: An Autobiography* (Boston: Houghton Mifflin, 1918), 380.

32. Ibid., 382.

33. R. P. Blackmur, *Henry Adams* (New York: Harcourt, Brace, Jovanovich, 1980), 244.

34. Jawaharlal Nehru, *The Discovery of India* (Delhi: Oxford University Press, 1985), 557.

35. Quoted in Itty Abraham, *The Making of the Indian Atomic Bomb: Science, Secrecy and the Postcolonial State* (New York: Zed, 1998), 28.

36. Ibid.

37. For further theoretical and ethnographic attention to the legal and political connections—and the political programs—that people strive to make across infrastructures (often in the face of their material dilapidation) and how these are crafted imaginatively, but also substantively, as claims on rights and assertions of social justice, see Stephen Collier, *Post-Soviet Social: Modernity, Neoliberalism, Biopolitics* (Princeton, NJ: Princeton University Press, 2011), and Antina

Von Schnitzler, "Traveling Technologies: Infrastructure, Ethical Regimes, and the Materiality of Politics in South Africa," *Cultural Anthropology* 28 (2013): 670–93. Earlier and indispensable accounts of infrastructure as a vehicle for imaginative connection and constructive cultural and political work include Apter, *Pan-African Nation*, and Fernando Coronil, *The Magical State: Nature, Money, and Modernity in Venezuela* (Chicago: University of Chicago Press, 1997).

38. Michael M. J. Fischer, "Four Genealogies for a Recombinant Anthropology of Science and Technology," *Current Anthropology* 22 (2007): 577.

39. Clifford Geertz, *Negara: The Theatre State in Nineteenth-Century Bali* (Princeton, NJ: Princeton University Press, 1980), 121.

40. Clifford Geertz, "Centers, Kings, and Charisma: Reflections on the Symbolics of Power," in *Local Knowledge: Further Essays in Interpretive Anthropology* (New York: Basic Books, 1983), 124.

41. See Giorgio Agamben, *The Kingdom and the Glory: For a Theological Genealogy of Economy and Government* (Stanford, CA: Stanford University Press, 1997), 168–77.

42. Ibid., 230.

43. Ibid., 232–53. At the outset of this discussion, Agamben draws on Marcel Mauss's study of prayer to make further, instructive, comparisons between modern political rituals and both Christian and Vedic precedents—which deserve closer and fuller study than I can grant them here.

44. See Edward Shils, "Charisma, Order, and Status," *American Sociological Review* 30, no. 2 (1965): 199–213. See also the analysis of charisma as urban "infra-power" in Thomas Blom Hansen and Oskar Verkaaik, "Urban Charisma: On Everyday Mythologies in the City," *Critique of Anthropology* 29 (2009): 5–26.

45. Hannah Arendt, *The Human Condition*, 2nd ed. (Chicago: University of Chicago Press, 1998), 204. Later on in this book, when we encounter Indian nationalists writing of electricity as the "life blood of the industrial nation" and advocating its further use as part of a broad-based liberating process, we might hear echoes of Arendt's use of the same vital vocabulary to talk about political power and its role in collective life more generally (see chapter 3). What matters more than these merely metaphorical resonances is that actors in India's political history sought to use technological powers to craft a relation to time and to each other—transforming their vitalist, energetic vocabulary of development directly into an ethical reflection (see the epigraph to this introduction).

46. Alfred North Whitehead, *Process and Reality*, corr. ed., ed. David Ray Griffin and Donald W. Sherburne (New York: Free Press, 1978), 18–19 and 28–29.

47. For this sense of "infrastructural" see Marshall Sahlins, "Infrastructuralism," *Critical Inquiry* 36 (2010): 371–85.

48. Charles Taylor, *A Secular Age* (Cambridge, MA: Harvard University Press, 2007), 174.

49. The "extended case" is, of course, a classic method in legal anthropology, in which disputes and conflicts are placed in their wider, meaning-granting, context; I take my inspiration for these "cases" of problematic electrification also, however, from cultural studies, where the idea of a "case" is derived from psychoanalytic and detective literature and is organized around a "problem-event that has animated some sort of judgment." Lauren Berlant, "On the Case," *Critical Inquiry* 33 (2007): 663; cf. Don Handelman, "The Extended Case," and Bruce Kapferer, "Situations, Crisis, and the Anthropology of the Concrete," in *The Manchester School: Practice and Ethnographic Praxis in Anthropology*, ed. T. M. S. Evens and Don Handelman (New York: Berghahn, 2006), 94–158.

50. Multiple regulations, and even certain legislative acts, are not included in this skeletal legislative history and only appear in the wider historical arc of this book out of sequence. For example the 1910 Indian Electricity Act, which expanded and replaced the minimal terms of the 1903, appears in this analysis only when it was interpreted by a court in 2008 (see chapter 5). This is a consequence of the anthropological method of the extended case study, which has different organizational imperatives than historiography does.

51. Don Handelman, *Models and Mirrors: Toward an Anthropology of Public Events*, 2nd ed. (New York: Berghahn, 1990).

52. I use "everyday" here not in order to refer to some separate, "popular" domain of action, but rather—following a distinction between two senses of the term clarified by Ritika Prasad in her recent social history of railways in India—to name the technologically saturated, quotidian, consciousness-forming medium of capitalist social relations and political experience in modern states. See Ritika Prasad, *Tracks of Change: Railways and Everyday Life in India* (Delhi: Cambridge University Press, 2015), 10. In the sense, used here, of a distinctively "modern" mediation of economic-political relations, the "everyday" is shaped as much by elite discourses and political rituals as it is by routine interactions, material installations, and the "popular" experience of them. The other sense of "everyday" is more empiricist and is commonly invoked in ethnographic studies. Work guided by this latter sense of the "everyday," by way of contrast to the methods adopted in this book, identifies routine interactions and popular experiences as sites of special ethnographic relevance and often as a domain *opposed to* (not constituted by) "high" (and *a priori* distorted) meanings and legitimating ideologies. This is the sense adopted by Chris Fuller and Véronique Bénéï in their ethnographic program for the study of "everyday state and society in India" and by David Arnold in his recent study of "everyday technology" in Indian modernity. See Arnold, *Everyday Technology*, and Christopher J. Fuller and Véronique Bénéï, eds., *Everyday State and Society in India* (London: Hurst, 2001).

53. I borrow this sense of "invention" from Roy Wagner—he describes the "invention" at stake in cultural performances as a condensation and correlation of meaning and action within a field of significations; invention in this sense is not reducible to the historicist notion of a singular, irreplicable moment of "invention." Rather, this kind of cultural invention can be endlessly repeated and reproduced, and can be aversive or reactive (counter-invention) as well as projective, while always remaining creative. Roy Wagner, *The Invention of Culture* (Chicago: University of Chicago Press, 1981).

54. Max Weber, *Economy and Society*, vol. 1 (New York: Bedminster Press, 1968), 330; cf. Kim Lane Scheppele, "Constitutional Ethnography: An Introduction," *Law and Society Review* 38 (2004): 395.

55. Sheila Jasanoff, *Designs on Nature: Science and Democracy in Europe and the United States* (Princeton, NJ: Princeton University Press, 2005).

56. This argument is most famously articulated by Partha Chatterjee in *Nationalist Thought and the Colonial World*.

57. Philip Oldenburg, *Big City Government in India: Councilor, Administrator, and Citizen in Delhi* (Tucson: University of Arizona Press, 1976), 349.

58. Ravi Sundaram, *Pirate Modernity: Delhi's Media Urbanism* (New York: Routledge, 2009).

59. Asok Mitra, *Delhi: Capital City* (Delhi: Thomson, 1970), 48.

1. The Machinery of Government

1. Stephen Wheeler, *History of the Delhi Coronation Durbar* (London: J. Murray, 1904), 57.

2. Plan of Durbar Amphitheater, in "A collection of proclamations, speeches, programs, maps, tickets and other material connected with the Delhi Durbar, 1903," General Reference Collection C.193.c.26, British Library.

3. This is the figure given by Wheeler; the official program (printed before the event) says thirty-seven thousand members of the Indian Army participated. See "Programme of Events at the Coronation Durbar, Delhi, 1903," General Reference Collection D.Q.7, British Library.

4. Wheeler, *Coronation Durbar*, 57.

5. Cohn, *Colonialism and Its Forms of Knowledge: The British in India* (Princeton, NJ: Princeton University Press, 1996), 4. Whatever the spectacular scale of the event, durbars remained

stately occasions, which required careful ritual technique to execute properly. Curzon wrote, much later in his life, that the regular durbars he held as viceroy, at the classical palace that was Government House in Calcutta, were the "most impressive" and the "most stately" of the official ceremonies "in which the representative of the Sovereign was frequently called upon to take part," and in his recollections he lavishes great descriptive detail on the minutiae of the protocol observed on these occasions. George Nathaniel Curzon, *British Government in India*, vol. 1 (London: Cassell, 1925), 237–39.

6. In its fostering of Indian princes as "traditional" rulers, the colonial state built upon a kind of feudal imaginary of ranked and graded difference that misrecognized the bases of both leadership and community in India and reified its misrecognitions into symbols of "native" power with which it then bedecked its own rationalized rule. See Bernard Cohn, "Representing Authority in Victorian India." See also Nicholas Dirks, *The Hollow Crown: Ethnohistory of an Indian Kingdom* (New York: Cambridge University Press, 1987), and Douglas Haynes, *Rhetoric and Ritual in Colonial India*. Colonialism in India was just one version of a widely shared European project of rule through ethnological knowledge about colonized societies' traditional patterns of kinship and power. As George Steinmetz has shown for German colonialism, especially in Asia and the Pacific, translation and misrecognition of indigenous categories, and performative enactments of the distorted result, were integral social and personal, even psychic, elements of the colonial project. Steinmetz, *The Devil's Handwriting: Precoloniality and the German Colonial State in Qingdao, Samoa, and Southwest Africa* (Chicago: University of Chicago Press, 2007).

7. It might be worth noting that Nicholas Dirks's description of the colonial state as a "theater state," his "impious appropriation" of Clifford Geertz's more complex use of the term to describe premodern Balinese courtly statecraft, is inapplicable here. Dirks's use of theatrical metaphors was tuned to reproduce the very oppositions between "merely" symbolic ritual practice and "real" material power that I am trying to show are unhelpful in understanding both the imperial ritualism that was a distinctive part of colonialism and political ritual more generally. See Dirks, *Hollow Crown*, 384. Dirks's later works show that colonial knowledge regimes and governmental practices had enormous cultural power, and the latter insight can be adapted to buttress my careful focus on the central ritual practices by which these colonial knowledges were ratified and affirmed. But still, as Anne Stoler has pointed out, "much of colonial studies over the past decade has worked from the shared assumption that the mastery of reason, rationality, and the inflated claims for Enlightenment principles have been at the political foundation of colonial regimes and should be at the center of critical histories of them." Rationalities and knowledges need to be understood not only as they were articulated, but also as they were mobilized affectively and even bodily—which is what ritual does. Stoler, *Along the Archival Grain: Epistemic Anxieties and Colonial Common Sense* (Princeton, NJ: Princeton University Press, 2009), 57.

8. Apter, *Pan-African Nation*, 181.

9. Curzon, *British Government*, 202. I borrow the phrase "permanent, protective" from Karuna Mantena's brilliant study of Henry Maine as the progenitor of the kind of high-Tory imperialism of which Curzon is an epigone. Mantena, *Alibis of Empire: Henry Maine and the Ends of Liberal Imperialism* (Princeton, NJ: Princeton University Press, 2010), 149. See also Mann, "Torchbearers."

10. Edward Said quotes a characteristic passage from a speech of Curzon's in which he said: "I sometimes like to picture to myself this great Imperial fabric as a huge structure like some Tennysonian 'Palace of Art,' of which the foundations are in this country, where they have been laid and maintained by British hands, but of which the |white| colonies are the pillars, and high above all floats the vastness of an Asiatic dome." In Edward Said, *Orientalism* (New York: Pantheon, 1978), 213.

11. George Nathaniel Curzon, *Lord Curzon in India: Being a Selection from His Speeches as Viceroy and Governor-General of India, 1898–1905*, ed. T. Raleigh (London: Macmillan, 1906), 576.

12. E.g., David Gilmour, *Curzon* (London: John Murray, 1994); David Dilks, *Curzon in India* (London: Rupert Hart-Davis, 1969); Lawrence John Lumley Dundas, Marquis of Zetland, *The Life of Lord Curzon: Being the Authorized Biography of George Nathaniel, Marquess Curzon of Kedleston, K.G., by the Rt. Hon. the Earl of Ronaldshay* (London: E. Benn, 1928).

13. One of the reasons that multiple railway stations had to be used for moving durbar attendees in and out of Delhi was to avoid the embarrassment of lower- and higher-ranking Indian princes arriving on adjacent platforms at the same time, with their competing ceremonial fanfares and salutes.

14. Curzon, *Lord Curzon in India*, 295.

15. Walter Bagehot, *The English Constitution* (London: Oxford University Press, 1928).

16. Curzon, *Lord Curzon in India*, 290.

17. Ibid.

18. For parliamentary mentions of the salt tax as a special iniquity of the overall policy of the government of India see *Hansard* (4th series) November 21, 1902, vol. 115, cc. 162, 165.

19. Viceroy in Council to Hamilton, 23 October 1902, IOR Eur Mss F 123/60.

20. Ibid.

21. Godley to Hamilton, 10 November 1902, IOR Eur Mss F 123/60.

22. Curzon to Hamilton, 12 November 1902, 17 November 1902, IOR Eur Mss F 123/60.

23. Zetland, *Life of Lord Curzon*, 224.

24. Curzon to HM Edward VII, 24 November 1902, IOR Eur Mss F 123/60.

25. See Alan Trevithick, "Some Structural and Sequential Aspects of the British Imperial Assemblages at Delhi: 1877–1911," *Modern Asian Studies* 24 (1990): 561–78, for useful summaries of the reaction in the Indian press.

26. Curzon, *Lord Curzon in India*, 289.

27. Quoted in Cohn, "Representing Authority," 192.

28. Curzon, *Lord Curzon in India*, 143.

29. Quoted in Dilks, *Curzon in India*, 254.

30. In addition to the published sources and specific archival references cited in the text, this account is based upon my examination of Curzon's own notes on the durbar, official pamphlets and ephemera, and other material held in the Curzon Collection, specifically in IOR Eur Mss F 111/274 and IOR Eur Mss 112/431, as well as photographs in the Curzon Collection (IOR Photo 430/9, 430/10, 430/11, 430/78, 430/79).

31. Wheeler, *Coronation Durbar*, 57.

32. Valentia Steer, *The Delhi Durbar, 1902–1903* (Madras: Higgenbotham, 1903), 19.

33. Wheeler, *Coronation Durbar*, 57.

34. Ibid., 152.

35. "The Delhi Durbar: A Retrospect," *Blackwood's Edinburgh Magazine*, March 1903, 318.

36. Mortimer Menpes and Dorothy Menpes, *The Durbar* (London: A. & C. Black, 1903), 44.

37. Gilmour, *Curzon*, 246.

38. Kashmir: Steer, *Delhi Durbar*, 21; Patiala: Wheeler, *Coronation Durbar*, 79.

39. Austin Cook, "An Account of the Delhi Durbar" (typescript), IOR Eur Mss D570.

40. Wheeler, *Coronation Durbar*, 62.

41. Gayatri Spivak, *A Critique of Postcolonial Reason: Toward a History of the Vanishing Present* (Cambridge, MA: Harvard University Press, 1999), 240.

42. Apter, *Pan-African Nation*, 177.

43. See Manu Bhagavan, "Demystifying the 'Ideal Progressive': Resistance through Mimicked Modernity in Princely Baroda, 1900–1913," *Modern Asian Studies* 35 (2001): 385–409.

44. Benedict Anderson, *Imagined Communities: Reflections on the Origins and Spread of Nationalism*, 2nd ed. (New York: Verso, 1991), 153.

45. Gupta, *Delhi between Two Empires*, 175.

46. Wilson to Municipal Committee, n.d. [March 1901], in CCO 1902, 58/79, "The Introduction of Electric Tramways and Light into Delhi Municipality."

47. Ibid.

48. "John Fleming & Co.," *The Cyclopedia of India* (Calcutta: Cyclopedia, 1907), 289–90.

49. According to incomplete membership lists cited by Gupta, *Delhi between Two Empires*, 235–38.

50. Haynes, *Rhetoric and Ritual*, 115–26.

51. Douglas to Fanshawe, 6 May 1901, CCO 1902 58/79.

52. Fanshawe to Judicial & General Secretary, Punjab, 8 May 1901, CCO 1902 58/79.

53. Connolly to Walker, 24 July 1902, including printed circulars of Government of Punjab Boards and Committees Department, Nos. 28–52, 1902, CCO 1902 58/79.

54. Connolly to Sir Thomas Higham, 6 June 1902 (printed circular), CCO 1902 58/79.

55. Higham to Connolly, 8 August 1902 (printed circular), CCO 1902 58/79.

56. Ibid.

57. Grey to Rivaz, 22 July 1902 (printed circular), CCO 1902 58/79.

58. As communicated by his secretary Connolly to Higham, 25 July 1902 (printed circular), CCO 1902 58/79.

59. Walker to Connolly, 30 June 1902, CCO 1902 58/79.

60. Douglas to Walker, 15 October 1902, CCO 1902 58/79.

61. Printed circular: Government of Punjab, Boards and Committees Department, June 1903, Nos. 20–26, CCO 1902 58/79.

62. "John Fleming & Co.," *Cyclopedia of India*, 290.

63. Parsons to Walker, 28 November 1904, CCO 1902 58/79.

64. Kettlewell to Grey, 7 July 1904 (emphasis added) (printed circular), CCO 1902 58/79.

65. *The Story of Bruce Peebles, 1866–1954* (National Library of Scotland 5.227). Further details here are culled from files in the National Records of Scotland relating to the liquidation of Bruce Peebles, particularly NAS GD282/13/81.

66. CCO 1907 186/63.

67. Hugh Tinker, *The Foundations of Local Self Government in India, Pakistan, and Burma* (London: Athlone, 1954), 60.

68. Stephen Legg, *Spaces of Colonialism: Delhi's Urban Governmentality* (Malden, MA: Wiley-Blackwell, 2007), 151.

69. Curzon, *Lord Curzon in India*, 550.

70. See accounts of the Viceroy's Camp, Coronation Durbar, Delhi, IOR Eur Mss F112/467.

71. George Peel, "At the Durbar," *The Cornhill Magazine* 14 (n.s.) (1903): 316.

72. Quoted in Gilmour, *Curzon*, 245.

2. Ritual Center and Divided City

1. Nirad C. Chaudhuri, *Thy Hand, Great Anarch!* (New Delhi: Times Books, 1987), 690.

2. M. K. Gandhi, *"Hind Swaraj" and Other Writings*, ed. Anthony J. Parel (New York: Cambridge University Press, 1997), 108–9.

3. Stanley Reed, *The King and Queen in India* (Bombay: Bennett, Coleman, 1912), 150.

4. Robert Grant Irving, *Indian Summer: Lutyens, Baker, and Imperial Delhi* (New Haven, CT: Yale University Press, 1981), 9–10.

5. George, King of Great Britain, *Speeches of His Majesty King George in India*, 2nd ed. (Madras: G. A. Natesan, 1912), 120.

6. For a useful summary of the rationale for partition and its reception in Bengali political society see J. H. Broomfield, *Elite Conflict in a Plural Society: Twentieth-Century Bengal* (Berkeley: University of California Press, 1968), 25–35. The "real" justification of the partition was openly

acknowledged among officials; Lord Hardinge wrote to a colleague in 1911 that while administrative convenience was the overt justification for the partition of Bengal, "the desire to aim a blow at the Bengalis overcame other considerations in giving effect to that laudable object"; quoted in F. A. Eustis and Z. H. Zaidi, "King, Viceroy, and Cabinet: The Modification of the Partition of Bengal, 1911," *History* 49 (1964): 179.

7. David Johnson, "Land Acquisition, Landlessness, and the Building of New Delhi," *Radical History Review* 108 (2010): 96.

8. R. E. Frykenberg, "The Coronation Durbar of 1911: Some Implications," in *Delhi through the Ages* (Delhi: Oxford University Press, 1986), 229.

9. Irving, *Indian Summer*, 11.

10. Reed's contemporaneous account in *The King and Queen in India* lauds the king's wisdom in pronouncing the change in this way and implies that the decision to associate the change in policy with the king's coronation was a political one—not because such an association would make whatever was decided more amenable to the "Indian mind," but because both changes were bound to be unpopular among the British in India. Using the king in this way achieved Britain's *Indian* political ends without leaving room for carping among officials and *British* politicians. This is, to me, at least as convincing an explanation of the negotiations around the boon as any intended effect on India's communities.

11. Reed, *King and Queen*, 166.

12. See Leo Coleman, "Ignorance and Government in British India: The Native Fetish of the Crown," in *Regimes of Ignorance: Anthropological Perspectives on the Production and Reproduction of Non-knowledge*, ed. Roy Dilley and Thomas G. Kirsch (New York: Berghahn, 2015), 159–87.

13. See Arthur Berriedale Keith, *The King and the Imperial Crown: The Powers and the Duties of His Majesty* (London: Longmans, Green, 1936); cf. Harold Laski, *Parliamentary Government in England* (New York: Viking, 1938).

14. "Report on Durbar Arrangements," DCO 1911 no. 44.

15. "Electrical Requirements of New Delhi," CCO, Commerce and Industry 1/1913B; "Proposed Supply of Electrical Energy from the Durbar Works Power Station," CCO, Commerce and Industry (Confidential Files) 1913; "Grants for the Government Electrical Plant," CCO, Financial 26/1913A.

16. Pollard-Lawsley to Chief Commissioner, n.d., CCO, Home 262/1914B; Chief Commissioner to Deputy Commissioner, 13 May 1915, CCO, Commerce and Industry 73/1915B.

17. Tilman Frasch, "Tracks in the City: Technology, Mobility, and Society in Colonial Rangoon and Singapore," *Modern Asian Studies* 46 (S01) (2012): 115. See also Meera Kosambi, *Bombay in Transition: The Growth and Social Ecology of a Colonial City* (Stockholm: Almqvist & Wiksell, 1986), for similar insights into the progressive history of Bombay's municipal improvements.

18. Delhi Electric Supply Enquiry (Pitkeathly) Committee Report, vol. 1 (New Delhi: Government of India, 1937), 10–11, IOR V/26/740/3.

19. Quoted in Abraham, *Indian Atomic Bomb*, 28.

20. CCO Industries B-37, 1929.

21. Delhi Electric Supply Enquiry Report, vol. 1, p. 5, IOR V/26/740/3.

22. Delhi Electric Supply Enquiry Report, vol. 2, p. 7, IOR V/26/740/4.

23. See David Johnson, "A British Empire for the Twentieth Century: The Inauguration of New Delhi, 1931," *Urban History* 35 (2008): 462–84.

24. Mann, "Torchbearers," 25.

25. James Boon has specified the substantial affinities between mechanical solidarity and charismatic legitimation, even in its bureaucratized forms; see James A. Boon, *Other Tribes, Other Scribes: Symbolic Anthropology in the Comparative Studies of Cultures, Histories, Religions, and Texts* (New York: Cambridge University Press, 1982), 81–82.

26. Nayantara Pothen, *Glittering Decades: New Delhi in Love and War* (New Delhi: Penguin, 2012), 62.

27. Irving, *Indian Summer*, 270.

28. The 1903 durbar grounds had been described by one British correspondent as appearing like an "exhalation rising suddenly from the plains of Delhi." Peel, "At the Durbar," 316.

29. Sten Nilsson, *The New Capitals of India, Pakistan, and Bangladesh*, trans. Elisabeth André-asson (Lund, Sweden: Studentlitteratur, 1973), 74.

30. "Notification dated Delhi, the 22nd March, 1939," CCO 1902 58/79.

31. Gandhi, *"Hind Swaraj,"* 33–36.

32. Ibid., 47.

33. Ruth Benedict, *Patterns of Culture* (New York: Houghton Mifflin, 1934), 55.

34. See Parama Roy, *Alimentary Tracts: Appetites, Aversions, and the Postcolonial* (Durham, NC: Duke University Press, 2010), for more on the intellectual affinities that linked Gandhi to wider currents of radical thought in the period.

35. Ajay Skaria, "Relinquishing Republican Democracy: Gandhi's Ramarajya," *Postcolonial Studies* 14 (2011): 224.

36. See Arindam Dutta, *The Bureaucracy of Beauty: Design in the Age of Its Global Reproducibility* (New York: Routledge, 2007), 249; Dutta directs attention to Gandhi's specific critique of the sumptuary requirements of the colonial state in his *Autobiography* (Ahmedabad: Navajivan, 1927), 212.

37. Gandhi, "Constructive Programme: Its Meaning and Place (1941, revised 1945)" in *"Hind Swaraj,"* 179–80.

38. Nehru's words, quoted ibid., 172.

39. See Gyan Prakash's discussion of archaism in nationalist discourse and his illuminating account of the relation between Gandhian and Nehruvian visions of the new nation, in *Another Reason: Science and the Imagination of Modern India* (Princeton, NJ: Princeton University Press, 1999), 86–120 and 216–21.

40. Nehru, *Discovery of India*, 29.

41. Sunila Kale quotes a 1934 article from *Harijan* in which Gandhi specifically identifies electrification with centralized "control of power" and says that the consequences of this technological control would be terrible. See Kale, *Electrifying India*, 28.

42. Milton Singer, *When a Great Tradition Modernizes: An Anthropological Approach to Indian Civilization* (New York: Praeger, 1972), 399–400.

43. At the very least, Gandhi granted Nehru a mass party to lead, and helped organize into unity a population at the head of which the latter could pursue his modernizing ambitions.

44. Gandhi, it might be said, independently arrived at something like Durkheim's understanding of society. In that understanding, "organic solidarity remains Durkheim's touchstone, and society is less 'corporate group' than 'divisioned whole.'" Boon, *Other Tribes*, 60. Obviously, it is significant for this conjuncture that both Durkheim and Gandhi were, in their ways, deeply "religious" thinkers, but inspired by very different traditions and with very different personal relations with those traditions; this is more than can be taken on in this context.

3. The Lifeblood of the Nation

1. See Johnson, "British Empire for the Twentieth Century."

2. Jawaharlal Nehru, *Glimpses of World History* (Delhi: Oxford University Press, 1989), 608.

3. Ibid., 848.

4. Indian National Congress, *Power and Fuel: Report of the Subcommittee of the National Planning Committee*, ed. K. T. Shah (Bombay: Vora, 1949), 71.

5. See *Collected Works of Meghnad Saha*, vol. 2, ed. Santimay Chatterjee (Calcutta: Orient Longman, 1986), 198.

6. Ibid., 190.

7. Delhi Electric Supply Enquiry (Pitkeathly) Committee Report, vol. 1 (New Delhi: Government of India, 1937), 41, IOR V/26/740/3.

8. Some 480,000 refugees were in Delhi by 1951. To give some sense of the scale of the changes in the fabric of the city, the government economist Asok Mitra later pointed to the fact that in the year and a half after independence, the interim government's Ministry of Works built "20 [refugee settlement] colonies, far flung as well as closely nestled . . . on 3,000 acres of land." In the following decade, the Ministry of Rehabilitation "provided Delhi with a total of 69,400 dwelling units sheltering 53 percent of the refugee population." Mitra, *Delhi*, 11.

9. Taylor, "Modern Social Imaginaries," in *Secular Age*, 159–211.

10. Taylor, *Secular Age*, 174–75.

11. See, e.g., Townsend Middleton, "Anxious Belongings: Anxiety and the Politics of Belonging in Subnationalist Darjeeling," *American Anthropologist* 115 (2013): 608–21.

12. Raghavan, *Asaf Ali's Memoirs*, 360. See Leslie Hannah, *Electricity before Nationalisation* (Baltimore: Johns Hopkins University Press, 1979), 313–28, for a detailed account of the political and economic bases of these shortages, both as a result of the transition away from wartime production and as part of the wider moves in the period to integrate and nationalize the UK electricity industry.

13. See Manu Bhagavan, *The Peacemakers: India and the Quest for One World* (New Delhi: HarperCollins, 2012), for the international dimension of this story and insight into the leading role played by women in the diplomacy and politics of the era. Not incidentally, Asaf Ali's wife, Aruna Asaf Ali, was a major figure in the underground nationalist resistance in the late-colonial period.

14. Raghavan, *Asaf Ali's Memoirs*, 360. This letter refers explicitly to the pre-partition question of whether the Congress Party should agree to a loose federal India, with power shared between Congress and the Muslim League. Asaf Ali uses the state of England's industrial economy as an argumentative lever with which to push Nehru to hold out for a strong centralized government that could foster the forces of a modern economy throughout undivided India.

15. Kale, *Electrifying India*, 31.

16. Sundaram, *Pirate Modernity*, 48–49.

17. S. K. D. Gupta, "Review of Legislation 1947: Eastern Countries, India—Central Legislature," *Journal of Comparative Legislation and International Law*, 3rd series, 31 (1949): 116.

18. Report of the Advisory Board on the Principles of Public Utility Electricity Supply Finance (22 November 1946), 10, IOR V/27/740/14.

19. Speech of N. V. Gadgil, reintroducing a more modest Electricity (Supply) Bill on December 6, 1947; *Constituent Assembly (Legislative) Debates*, vol. 2 (Delhi: Government of India, 1948): 1398–99.

20. *Constituent Assembly (Legislative) Debates*, 2:1419.

21. Ibid., 2:1465–66.

22. Kale, *Electrifying India*, 34.

23. It may not be incidental, in this regard, that many Indian nationalist elites gained their first experience of practical government in municipal institutions.

24. Dhirendranath Sen, *From Raj to Swaraj* (Calcutta: Vidyodaya, 1954), 262.

25. See note 8 above.

26. See, e.g., *Constituent Assembly Debates*, vol. 9 (New Delhi: Government of India, 1967), 78.

27. Granville Austin, *The Indian Constitution: Cornerstone of a Nation* (Oxford: Clarendon, 1966), 44.

28. *Constituent Assembly Debates*, 9:100 (2 August 1949).

29. Ibid., 9:76–77 (1 August 1949).

30. Ibid., 9:78 (1 August 1949).

31. Ibid., 9:98 (2 August 1949).

32. Ibid., 9:96 (2 August 1949).

33. Ibid., 9:97 (2 August 1949).

34. For a handy survey of administrative and governmental arrangements for Delhi—which goes into much more detail than I can here, including brief periods of home rule in the 1950s—see Isabelle Milbert, "Law, Urban Policies, and the Role of Intermediaries in Delhi," in *New Forms of Urban Governance in India: Shifts, Models, Networks, and Contestations*, ed. I.S.A. Baud and J. de Wit (New Delhi: Sage, 2008), 184–92.

35. See Emma Tarlo, *Unsettling Memories: Narratives of the Emergency in Delhi* (Berkeley: University of California Press, 2003), for a ethnographic account of the period of the emergency, when local and central politics were merged and routed through the DDA and central institutions of urban development, with devastating effect. Ultimately, however, "home rule" would reappear as a political slogan and effective demand in Delhi in the 1990s—as I discuss in chapter 5—and the resulting creation of a separate Delhi State would inaugurate a new phase of political life in the city.

36. See, especially, Srirupa Roy, *Beyond Belief: India and the Politics of Post-colonial Nationalism* (Durham, NC: Duke University Press, 2007); cf., Ronald Inden, "Embodying God: From Imperial Progresses to National Progress in India," *Economy and Society* 24 (1995): 245–78.

37. Nehru to Mayer, 14 May 1947, Albert A. Mayer Papers (hereafter, AMP), box 8, file 1, University of Chicago Library.

38. See Matthew Hull, "Communities of Place, Not Kind: American Technologies of Neighborhood in Postcolonial Delhi," *Comparative Studies in Society and History* 53 (2011): 757–90. Hull quotes Douglas Ensminger, part of the later Ford Foundation master-planning team in Delhi, to the effect that Mayer had a "warm relationship" with Nehru.

39. Mayer to Nehru, 23 October 1950. AMP, box 8, file 5.

40. See Committee on Emotional Integration, *Report* (New Delhi: Ministry of Education, 1962).

41. Mayer to Nehru, 23 July 1953, italics his. AMP, box 8, file 4.

42. Nehru to Mayer, 29 July 1953, AMP, box 8, file 5.

43. Paul Rabinow, drawing from Foucault, has described the contemporary as defined by a "moving ratio" between techne, ethics, and power, and as a "problem-space" in which science, technology, and political and ethical reflection are intertwined and mutually informative as they meet specific demands for knowledge and change. In principle, past moments in political history can also be treated as just such spaces, by attending to the intellectual and ethical labor of actors within them. Rabinow, *Marking Time: On the Anthropology of the Contemporary* (Princeton, NJ: Princeton University Press, 2007), 2.

44. Chatterjee, *Nationalist Thought*, 147.

45. Ibid., 169.

46. See ibid., 147. For an incisive critique of Chatterjee's position see Sugata Bose, "Instruments and Idioms of Colonial and National Development: India's Historical Experience in Comparative Perspective," in *International Development and the Social Sciences: Essays on the History and Politics of Knowledge*, ed. F. Cooper and R. Packard (Berkeley: University of California Press, 1997), 45–63.

47. Hannah Arendt, *On Revolution* (1963; New York: Penguin, 2005).

48. Uday Singh Mehta, "The Social Question and the Problem of History after Empire," in *Lineages of Empire: The Historical Roots of British Imperial Thought*, ed. Duncan Kelly (New York: Oxford University Press, 2009), 34.

49. Arendt, *On Revolution*, 102.

50. Mehta, "Social Question," 49.

51. Mehta, "Social Question," 32; see Arendt, *On Revolution*, 85.

52. Ibid., 55–56.

53. See Jonathan Parry, "Nehru's Dream and the Village 'Waiting Room': Long-Distance Migrants to an Indian Steel Town," *Contributions to Indian Sociology* (n.s.) 37 (2003): 219–49; and "The Sacrifices of Modernity in a Soviet-Built Steel Town in Modern India," in *On the Margins of Religion*, ed. Frances Pine and João de Pina-Cabral (Oxford: Berghahn, 2008), 233–62.

54. Ved Mehta, *Portrait of India* (New York: Farrar, Straus & Giroux, 1970), 285.

55. Ibid., 299.

56. Ibid., 288.

57. This working out of relations between stately meanings and citizen identity is integral to what Carol Greenhouse has called "empirical citizenship." Greenhouse, *The Paradox of Relevance: Ethnography and Citizenship in the United States* (Philadelphia: University of Pennsylvania Press, 2011), 262.

58. As Gyan Prakash has argued, through the rationalist pursuit by elites of a form of knowledge-power that would serve India's national needs, the very instruments and institutions of state power in India were "surreptitiously incorporated in the inner domain" of Indian values and became part of a popular sense of India as a modern nation. Prakash, *Another Reason*, 230.

59. Nehru, *Discovery*, 557.

60. Roy, *Beyond Belief*, 111.

4. Broadcast Mantras

1. V. S. Naipaul, *India: A Wounded Civilization* (New York: Knopf, 1977). Reviewing other work by Naipaul in the *New York Review of Books* (June 12, 1980), Joan Didion caught precisely both the biliousness and the excoriating, demythologizing impulses that course through Naipaul's writing. "He persists in translating underdeveloped into underequipped, undereducated, undone by imported magic and borrowed images, metaphors, fantasies and applauded lies, fairy tales. He posits what has been the controlling historical trope of our time—the familiar image of the new world emerging from the rot of the old, the free state from the chrysalis of colonial decay—as a fairy tale, a rhetorical commodity, and his contempt for those who trade in it is almost total."

2. Clifford Geertz, *The Social History of an Indonesian Town* (Cambridge, MA: MIT Press, 1965), 12.

3. Naipaul, *India*, 87–88, paragraphing altered.

4. Ibid., 86.

5. Stuart Corbridge and John Harriss, *Reinventing India: Liberalization, Hindu Nationalism, and Popular Democracy* (Malden, MA: Blackwell, 2000), 192.

6. Saurabh Dube, "Lost and Found: Villages between Anthropology and History," in *Village Matters: Relocating Villages in the Contemporary Anthropology of India*, ed. Diane Mines and Nicolas Yazgi (New Delhi: Oxford University Press, 2010), 36.

7. André Béteille, *Society and Politics in India: Essays in a Comparative Perspective* (London: Athlone, 1991), 34; cf. Arjun Appadurai, "Review: Is Homo Hierarchicus?," *American Ethnologist* 13 (1986): 745–61.

8. See McKim Marriott, "Little Communities in an Indigenous Civilization," in *Village India: Studies in the Little Community*, ed. McKim Marriott (Chicago: University of Chicago Press, 1955), 171–222; and Louis Dumont and David Pocock, "Village Studies," *Contributions to Indian Sociology* 1 (1957): 23–41. In a programmatic review essay published in 1963, Marriott observed that no ethnographer of contemporary India could afford to leave "time and the city" out of the account. See McKim Marriott, review of *Gopalpur: An Indian Village*, by Alan Beals, *American Anthropologist* 65 (1963): 1369.

9. Diane P. Mines and Nicolas Yazgi, "Introduction: Do Villages Matter?," in *Village Matters: Relocating Villages in the Contemporary Anthropology of India* (New Delhi: Oxford University Press, 2010), 8.

10. For general intellectual background see Nils Gilman, *Mandarins of the Future: Modernization Theory in Cold-War America* (Baltimore: Johns Hopkins University Press, 2007). For the specific "liberal ideology" of Redfield's Comparative Civilizations projects see Andrew Sartori, "Robert Redfield's Comparative Civilizations Project and the Political Imagination of Postwar America," *positions* 6 (1998): 33–65, and Dipesh Chakrabarty, "Reconstructing Liberalism? Notes toward a Conversation between Area Studies and Diasporic Studies," *Public Culture* 10 (1998): 457–81. An early turn against modernization theory was offered on the basis of Indian research, but staying within its own Weberian terms, by Milton Singer, *When a Great Tradition Modernizes*.

11. M. N. Srinivas, *The Remembered Village* (Berkeley: University of California Press, 1976).

12. See M. N. Srinivas, "The Dominant Caste in Rampura," *American Anthropologist*, n.s., 61 (1959): 1–16.

13. Srinivas, *Remembered Village*, 244.

14. M. N. Srinivas, "A Note on Mr. Goheen's Note: Forum on India's Cultural Values and Economic Development," *Economic Development and Cultural Change* 7 (1958): 5; cf., *Remembered Village*, 63.

15. Srinivas, *Remembered Village*, 245.

16. See, in particular, Sartori, "Robert Redfield."

17. Marriott, "Little Communities," 173.

18. Ibid., 198–99.

19. David Mandelbaum, "The World and the World View of the Kota," in Marriott, *Village India*, 250.

20. Dumont and Pocock, "Village Studies," 36.

21. Ibid., 39.

22. Thomas Blom Hansen, *The Saffron Wave: Democracy and Hindu Nationalism in Modern India* (Princeton, NJ: Princeton University Press, 1999).

23. See Corbridge and Harris, *Reinventing India*, 192, and note 5, above.

24. Srinivas, *Remembered Village*, 238; he comments a bit later (p. 254) that "a long-range view of the changes which had occurred in Rampura revealed the crucial role played by the state," stretching back to the irrigation works that made the village lands productive in the eighteenth century, and all the way to the then-new electrification of the village.

25. Mitra, *Delhi*, 48.

26. Ibid., 4–5.

5. The Life of Property

1. The so-called rise of the middle classes, and their specifically political impact as a new interest group and dominant ideological force in the post-liberalization Indian state, have attracted a great deal of attention (I discuss relevant works throughout this chapter). A comprehensive sociological overview of the emergent phenomenon, contemporary with my fieldwork, is Leela Fernandes, *India's New Middle Class: Democratic Politics in an Era of Economic Reform* (Minneapolis: University of Minnesota Press, 2006). More recently, Erica Bornstein and Aradhana Sharma have nicely elucidated, on the basis of contemporary fieldwork with middle-class social movements in Delhi, what they call the distinctively "techno-moral" politics of reform and its transformative effects on the Indian state. See Bornstein and Sharma, "The Righteous and the Rightful: The Technomoral Politics of NGOs, Social Movements, and the State in India," *American Ethnologist* 43 (2016): 76–90.

2. For contemporary critical definitions and studies of this so-called "new politics" as distinct from the clientalistic and state-centered old politics of developmental India see Partha Chatterjee, *Politics of the Governed: Reflections on Popular Politics in Most of the World* (New York: Columbia University Press, 2004), and John Harriss, *Power Matters: Essays on Institutions, Politics, and Society in India* (New Delhi: Oxford University Press, 2006), esp. 256–74.

3. Further ethnographic insights into the cultural politics of neoliberal urban reform in India can be found in Simanti Dasgupta, *Bits of Belonging: Information Technology, Water, and Neoliberal Governance in India* (Philadelphia: Temple University Press, 2015); and the essays in Swapna Bannerjee-Guha, ed., *Accumulation by Dispossession: Transformative Cities in the New Global Order* (New Delhi: Sage, 2010). Cf., for a picture of neoliberal electricity reforms in the United States, David Hess, "Electricity Transformed: Neoliberalism and Local Energy in the United States," *Antipode* 43 (2011): 1056–77. See Collier, *Post-Soviet Social*, for a *longue durée* analysis of these developments, particularly in post-Soviet Russia, as they reveal global changes in the rationalities of governance.

4. See, e.g., Saskia Sassen, *Territory, Authority, Rights: From Medieval to Global Assemblages* (Princeton, NJ: Princeton University Press, 2006).

5. For a theoretical argument about how work with legal technicalities can create "private constitutional moments," affecting larger-scale political interdependencies and even shaping a kind of solidarity around common interests, see Annelise Riles, *Collateral Knowledge: Legal Reasoning in Global Financial Markets* (Chicago: University of Chicago Press, 2011), esp. chap. 4; the phrase "private constitutional moments" is on p. 177.

6. See D. Asher Ghertner, "Calculating without Numbers: Aesthetic Governmentality in Delhi's Slums," *Economy and Society* 39 (2010): 185–217, and "Nuisance Talk and the Propriety of Property: Middle-Class Discourses of a Slum-Free Delhi," *Antipode* 44 (2012): 1161–87; Rohan Kalyan, "The Magician's Ghetto: Moving Slums and Everyday Life in a Postcolonial City," *Theory, Culture & Society* 31 (2014): 49–73; and Sanjay Srivastava, "Urban Spaces, Disney-Divinity, and Moral Middle Classes in Delhi," *Economic and Political Weekly* 44 (2009): 338–45. Partha Chatterjee's classic analysis, which is foundational for all these studies, has pointed to the "disengagement of the middle-classes from the hurly-burly of urban politics" and their deliberate self-restriction "to the strictly non-political world of the NGOs" as a defining feature of urban political life in India. The political mobilizations that emerged, after my fieldwork, over corruption and reform in Delhi seem to demonstrate, however, that a partial convergence is under way between the forms of "civic" association and NGO activism that Chatterjee dismisses as "non-political" and the party-political and state structures that engage with what he has defined as "political society." This observation is borne out by the recent theorization of NGO politics and the rise of the Aam Aadmi Party by Bornstein and Sharma (cited above). See Chatterjee, *Politics of the Governed*, 142.

7. William Mazzarella, "Internet X-Ray: E-Governance, Transparency, and the Politics of Immediation in India." *Public Culture* 18 (2006): 497.

8. See chaps. 1 and 2, above, and for more on the moral and spatial boundaries characteristic of colonial governmentality, see also Legg, *Spaces of Colonialism*.

9. John L. Comaroff and Jean Comaroff, "Reflections on the Anthropology of Law, Governance, and Sovereignty," in *Rules of Law and Laws of Ruling: On the Governance of Law*, ed. Franz von Benda-Beckmann, Keebet von Benda-Beckmann, and Julia Eckert (Burlington, VT: Ashgate, 2009), 31–33.

10. See Sunila S. Kale, "Current Reforms: The Politics of Policy Change in India's Electricity Sector," *Pacific Affairs* 77 (2004): 481. As should be evident, organized labor was absent from the contexts in which I conducted my fieldwork, and indeed it was all but invisible in the public conversation over privatization in Delhi in 2005. For more on the wider processes of reform in India see Joël Ruet, *Privatising Power Cuts? Ownership and Reform of the State Electricity Boards in India*

(New Delhi: Academic Foundation / Centre des Sciences Humaines, 2005), and Kale, *Electrifying India*.

11. In Rohit Saran, "How We Live: Census of India Household Survey," *India Today*, international edition, July 28, 2003, 10–17.

12. William Mazzarella, *Shoveling Smoke: Advertising and Globalization in Contemporary India* (Durham, NC: Duke University Press, 2003), 70. Italics his.

13. The Sheila Dikshit government was easily reelected immediately following privatization in 2002 (though before it was implemented), and some people I talked to cited the Congress government's electoral success as a sign that there was neither opposition to, nor discontent with, the power privatization in Delhi—just at the moment that post-privatization activism was taking shape.

14. See the Dikshit government's own narrative of the reforms, available at http://delhigovt. nic.in/newdelhi/power.asp.

15. The point that "illegality" is just as much a feature of upper-class life in Delhi is well made, in regard to water and illegal tube-wells, by Yaffa Truelove and Emma Mawdsley, "Discourses of Citizenship and Criminality in Clean, Green Delhi," in *A Companion to the Anthropology of India*, ed. Isabelle Clark-Decès (Malden, MA: Blackwell, 2011), 407–25.

16. Oldenburg, *Big City Government*, 349. Italics his.

17. The act empowers private utility companies to investigate civil complaints of dishonest use and assess damages on their own account, while setting up separate "electricity courts" for criminal infractions. See J. L. Bajaj and Anish De, "Electricity Act, 2003, and the Emerging Regulatory Challenges," *International Journal of Regulation and Governance* 4 (2004): 51–70, and, for more detailed scrutiny of the criminal provisions of the act, Subhes C. Battacharyya, "Review of the Electricity Act 2003 of India," *Dundee Center for Energy, Petroleum and Mineral Law and Policy Online Journal* 14 (2003), http://www.dundee.ac.uk/cepmlp/journal/html/volume14/article14_4. pdf. The act and its amendments are available online at http://www.powermin.nic.in/acts_notifi cation/electricity_act2003/preliminary.htm.

18. Nishtha Chugh, "Delhi's Power Woes," in *Frontline*, August 14, 2004, http://www.front line.in/static/html/fl2117/stories/20040827001005200.htm.

19. This was the consensus at a World Bank–funded meeting I attended in March 2006 at the International Development Fund in New Delhi. The research presented there was later published as Santhakumar, *Analysing Social Opposition*.

20. For further, detailed comparisons of bhagidari with other projects of decentralization in India see Joël Ruet and Stéphanie Tawa Lama-Rewal, eds., *Governing India's Metropolises* (New Delhi: Routledge, 2009), and, on Local Area Management in Mumbai, Marie-Hélène Zérah, "Reconfiguring Power Relations: Policies toward Urban Services in Mumbai," in *Accumulation by Dispossession*, ed. Swapna Bannerjee-Guha, 151–68. Cf. Milbert, "Law, Urban Policies, and the Role of Intermediaries in Delhi."

21. Lalitha Kamath and M. Vijayabaskar, "Limits and Possibilities of Middle Class Associations as Urban Collective Actors," *Economic and Political Weekly* 44 (2009): 368–76; see the unfavorable comparisons of bhagidari with other methods of participatory governance in K.C. Sivaramakrishnan, ed. *People's Participation in Urban Governance* (New Delhi: Institute of Social Sciences, 2006).

22. "Bhagidari or Jagirdari?," *Times of India*, Delhi edition, June 14, 2003, http://timesofindia. indiatimes.com/articleshow/22450.cms.

23. The rhetorical force of this question rests on a simple but effective pun on the Persian root "-dar"—status of control over, lordship—which *baghidari* (*bagh*=share, portion) shares with terms like "landowner" (*zamindar*) or precolonial noble titles such as *jagirdar* (which can be rendered as "estate holder"). Popular campaigns against zamindari and jagirdari were part of the wider nationalist movement.

24. See, further, Anne Waldrop, "Gating and Class Relations: The Case of a New Delhi 'Colony,'" *City and Society* 16, no. 2 (2004): 93–116, and Srivastava, "Urban Spaces, Disney-Divinity."

25. A 3 percent margin of error is the maximum allowed by law, under provisions that have not changed since the Indian Electricity Act of 1910. The RWA activist's knowledge of and use of legally established standards in his complaint is characteristic of the groups' "civil" appeals to law and order.

26. Aman Sethi, "A Shock to Reformers / The Price of Reforms," *Frontline*, September 10–23, 2003, http://www.frontline.in/static/html/fl2219/stories/20050923002703700.htm.

27. Rajeswari Sunder Rajan thus asks "whether the state has been so 'weakened that it cannot implement its own policies,' so that 'in exasperation, or for legitimacy,' it is forced 'to urge the masses to organize themselves against the state'" (internal citations omitted). See Sunder Rajan, *The Scandal of the State: Women, Law, and Citizenship in Postcolonial India* (Durham, NC: Duke University Press, 2003), 35.

28. This explanation is drawn from a now defunct "meter awareness" website run by the Delhi Electricity Regulatory Commission, on file with the author.

29. Richi Verma and Vaishali Saraswat, "Discoms' Lessons: Real or Fake?," *Times of India*, November 19, 2005, http://timesofindia.indiatimes.com/articleshow/1300793.cms. See also the Regulatory Commission's "Report of the Meter Testing Drive," on file with the author.

30. Delhi Electric Regulatory Commission "Meter Awareness," on file with the author.

31. "Demand to Disband DERC," *Hindu Online*, November 21, 2005, http://hinduonnet.com/2005/11/21/stories/2005112111180400.htm.

32. For parallel insights into the overlap of moral, aesthetic, and legal categories at stake in this activism, also drawn from fieldwork in Delhi, see Bornstein and Sharma, "Righteous and the Rightful," and Webb, "Short Circuits."

33. *Jindal vs. BSES Rajdhani* 126 (2006) DLT 49 (at paragraph 19) (all legal cases consulted via indiankanoon.com).

34. *Jindal vs. BSES Rajdhani* (2008) 1 SCC 341 (at paragraph 40).

35. To be clear, this case involved provisions of the 1910 Electricity Act that still regulated private concerns and had not been superseded by later legislation—hence the court's use of the term "licensee," and providing a further link back to colonial proprieties. Since BSES had taken over distribution in parts of Delhi prior to the enactment of the national Electricity Act in 2003, that law's provisions did not yet govern the relation between discoms and customers in Delhi.

36. *Jindal vs. BSES* (2008) 1 SCC 341, at paragraph 43.

37. The court cites its own earlier decision in *Maharashtra vs. Desai* (2003) 4 SCC 601, in which it quotes from Bennion's "Statutory Interpretation" to the effect that "in its application on any day, the language of the Act though necessarily embedded in its own time, is nevertheless to be construed in accordance with the need to treat it as a current law."

38. Chris Garces, "The Cross Politics of Ecuador's Penal State," *Cultural Anthropology* 25 (2010): 459–96.

39. Karl Marx, "Economic and Philosophical Manuscripts," in *Early Writings*, trans. Rodney Livingstone and Gregor Benton (New York: Vintage, 1975), 351 (and see the epigraph to this chapter); cf. Lauren Berlant, *Cruel Optimism* (Durham, NC: Duke University Press, 2011), 31, for a reading of this passage in terms of an affective economy of property.

40. Riles, *Collateral Knowledge*, 177; see note 5, above.

41. Carol J. Greenhouse, introduction in *Ethnographies of Neoliberalism*, ed. C. J. Greenhouse (Philadelphia: University of Pennsylvania Press, 2010), 3.

42. Mariana Valverde, "Seeing Like a City: The Dialectic of Modern and Premodern Ways of Seeing in Urban Governance," *Law & Society Review* 45 (2011): 306.

43. In his own attempt to untangle, analytically, the threads of class, status, and formal political belonging, Max Weber comments that "the road . . . from [a] purely conventional [status]

situation to legal privilege, positive or negative, is easily travelled as soon as a certain stratification of the social order has in fact been 'lived in' and has achieved stability by virtue of a stable distribution of economic power." Weber himself makes this point to stress that if one wishes to analyze modern forms of state power, one cannot investigate only the technical or legal instruments of the state, nor just the power and violence it wields, nor simply describe the spiritual garb it adopts, but rather should seek to understand the intense and mutual exchanges and transformations between these. Or, to make this same point another way, it is just what that "living in" consists in, beyond simple economic distribution, how it is organized and justified, that is the key ethnographic question, and where the important rites of participation and forceful exclusions must be specified. See Max Weber, "Class, Status, Party," in *From Max Weber: Essays in Sociology*, trans. H. H. Geerth and C. Wright Mills (New York: Oxford University Press, 1946), 188.

6. A Model Colony

1. Kaviraj, "Enchantment," 291.

2. It may be apposite to note here that Gandhi, disparaging the notion that India needed imported machinery in order to be free, writes in *Hind Swaraj* about India's subordinated position in the colonial economy that even "our gods are made in Germany." Gandhi, *"Hind Swaraj,"* 107.

3. Geertz, *Negara*, 121.

4. Greenhouse, *Paradox of Relevance*, 262.

5. There is a distinctive methodological resemblance that remains to be explored (on some other occasion) between Geertz's project of understanding archaic forms of state power by examining their ritual concentrations as well as their material distributions of power, and Greenhouse's forward-looking one of seeking empirical articulations of the state (as a source of the power and coordination needed and desired for the realization of democracy) in contemporary ethnographic accounts of United States urban culture. The affinity I find between these two distinct scholarly projects is, in part, a product of their similar ethnographic attention to rituals or performances that deliberately draw on relations and associations far beyond those given in a local context or ramified in a material network. Their shared anthropological interest is in cultural practices that reticulate relations—divide *and* integrate, or "weave"—to achieve extensive and multiple co-implications across apparent divisions and boundaries. They both help point the way to a broader anthropological understanding that links state power to its ritual iterations (or rather, seeks to grasp the mobile relations between power and ritual) while not reducing either to reflexes of material organization.

Conclusion

1. The very theoretically sophisticated artists of Raqs refer, with this piece, to Jacques Rancière's work on the "distribution of the sensible" as an integral aspect of class politics, while their laborer's diary recalls the nineteenth-century labor education movements that Rancière studied (in earlier work) to learn how workers staked a claim to participation from within, and through the employment of, industrial urbanism's own unequal economy of signs and values. See, e.g., Rancière, *The Politics of Aesthetics: The Distribution of the Sensible*, trans. Gabriel Rockhill (New York: Continuum, 2004).

2. Hocart, *Kings and Councillors*, 230–31; see the discussion of this passage from Hocart in the preface, above.

3. Rajeswari Sunder Rajan observes, acutely, that the "disjuncture" between constitutional provisions (with their liberal guarantees of rights) and material inequalities in the application of state power (either in the distribution of benefits or access to justice) is not the same thing as, nor is it as easily critiqued as, the "(inevitable) gap between promise and practice." Sunder Rajan,

Scandal of the State, 6–7. Disjunctures can be productive, at times even protective, and moreover can provide the space for imaginative reworking of failed, impossible, or lapsed promises.

4. The phrase is Daniel Miller's, and in full it reads: "In dialectical thought, proper materialism is one that recognizes the irreducible relation of culture." See Miller, introduction in *Materiality*, ed. Daniel Miller (Durham, NC: Duke University Press, 2005), 17.

5. Taylor, *Secular Age*, 174.

6. Singer, *When a Great Tradition Modernizes*, 399–400. For a broader argument about the importance of such ethical symbols in nationalist thought in India see Ananya Vajpeyi, *Righteous Republic: The Political Foundations of Modern India* (Cambridge, MA: Harvard University Press, 2012).

7. See the discussion in the introduction of charisma as a quality of roles and institutions, and Shils, "Charisma, Order, and Status."

8. I am not broaching here the historical questions raised by the translation of "community" from government usages to popular vernaculars in India and the effects this may have had but rather referring only to the term's English-language usage in current debates. One might note, however, that tracing translations and cross-appropriations of terms such as "public" and "community," both conceptually and practically, into and out of the local languages of popular politics in India has provided broad theoretical inspiration to scholars of Indian politics. See, for a recent review, J. Barton Scott and Brannon D. Ingram, "What Is a Public? Notes from South Asia," *South Asia: Journal of South Asian Studies* 38 (2015): 357–70.

9. I am thinking of the chapters "Two Histories of Capital" and "Nation and Imagination" in Dipesh Chakrabarty's *Provincializing Europe: Postcolonial Thought and Historical Difference*, 2nd ed. (Princeton, NJ: Princeton University Press, 2008); "Adda," also in that volume, provides the beginnings of a reconstruction of community of the sort I sketch in what follows.

10. See Saurabh Dube, "Presence of Europe: An Interview with Dipesh Chakrabarty," *South Atlantic Quarterly* 101 (2002): 860.

11. See Leela Gandhi, *Affective Communities: Anticolonial Thought, Fin-de-Siècle Radicalism, and the Politics of Friendship* (Durham, NC: Duke University Press, 2006), 23–26, where she discusses Blanchot's *Unavowable Community*. See also Roberto Esposito's *Communitas: The Origin and Destiny of Community*, trans. Timothy Campbell (Stanford, CA: Stanford University Press, 2010), especially the appendix where he outlines his debt to Nancy. I am grateful to Philip Armstrong for pointing me to this literature and sharing his own thoughts with me. See Philip Armstrong, *Reticulations: Jean-Luc Nancy and the Networks of the Political* (Minneapolis: University of Minnesota Press, 2009). See further Gerald Creed's incisive observation that community has operated in anthropology and sociology as an "aspiration envisioned as an entity," and his attempt to refigure community in more "divisioned" ways, in *The Seductions of Community: Emancipations, Oppressions, Quandaries* (Santa Fe, NM: School of Advanced Research Press, 2006), 22.

12. L. Gandhi, *Affective Communities*, 26; Chakrabarty, "Adda," in *Provincializing Europe*.

13. Numa Denis Fustel de Coulanges, *The Ancient City: A Study on the Religion, Laws, and Institutions of Greece and Rome* (Baltimore: Johns Hopkins University Press, 1980), 124.

BIBLIOGRAPHY

Archival Sources

British Library, London
    India Office Records (IOR) and select materials
Delhi State Department of Archives, Delhi, India
    Chief Commissioner's Records (CCO)
    Deputy Commissioner's Records (DCO)
University of Chicago Library, Chicago
    Mayer, Albert A., Papers on India
National Records of Scotland, Edinburgh
    Bruce Peebles & Co., papers relating to liquidation and reconstruction, 1906–1908

Government Reports and Publications

Delhi Electricity Regulatory Commission. *Report of the Meter Testing Drive for Electronic Meters* (in association with the Bureau of Indian Standards and the Central Power Research Institute). N.d. www.derc.gov.in/ElectronicMeters.
Government of India. *Delhi Electric Supply Enquiry (Pitkeathly) Committee, Volume 1: Report*, and *Volume 2: Representations Received*, 1937. Archived material, India Office Records, British Library, London: V/26/740/3 and V/26/740/4.
India. Committee on Emotional Integration. *Report*. Delhi: Ministry of Education, 1962.
India. Constituent Assembly. *Constituent Assembly (Legislative) Debates*. Vol. 2. Delhi: Government of India, 1948.

India. Constituent Assembly. *Debates*. Vol. 9. Delhi: Lok Sabha Secretariat, 1950 (1966–1967 reprint).

Indian National Congress, National Planning Committee. *Power and Fuel: Report of the Sub-Committee*. Edited by K. T. Shah. Bombay: Vora, 1949.

Books and Articles

Abélès, Marc. *The Politics of Survival*. Translated by Julie Kleinman. Durham, NC: Duke University Press, 2010.

Abraham, Itty. *The Making of the Indian Atomic Bomb: Science, Secrecy and the Postcolonial State*. New York: Zed Books, 1998.

Adams, Henry. *The Education of Henry Adams: An Autobiography*. Boston: Houghton Mifflin, 1918.

Agamben, Giorgio. *The Kingdom and the Glory: For a Theological Genealogy of Economy and Government*. Translated by Lorenzo Chiesa, with Mateo Mandarini. Stanford, CA: Stanford University Press, 2011.

Anderson, Benedict. *Imagined Communities: Reflections on the Origins and Spread of Nationalism*. 2nd ed. New York: Verso, 1991.

Appadurai, Arjun. "Review: Is Homo Hierarchicus?" *American Ethnologist* 13 (1986): 745–61.

Apter, Andrew. *The Pan-African Nation: Oil and the Spectacle of Culture in Nigeria*. Chicago: University of Chicago Press, 2005.

Arendt, Hannah. *The Human Condition*. 2nd ed. Chicago: University of Chicago Press, 1998.

———. *On Revolution*. New York: Penguin, 2005. First published 1963 by Viking.

Armstrong, Philip. *Reticulations: Jean-Luc Nancy and the Networks of the Political*. Minneapolis: University of Minnesota Press, 2009.

Arnold, David. *Everyday Technology: Machines and the Making of Indian Modernity*. Chicago: University of Chicago Press, 2013.

Austin, Granville. *The Indian Constitution: Cornerstone of a Nation*. Oxford: Clarendon, 1966.

Bagehot, Walter. *The English Constitution*. London: Oxford University Press, 1928.

Bajaj, J. L., and Anish De. "Electricity Act, 2003, and the Emerging Regulatory Challenges." *International Journal of Regulation and Governance* 4 (2004): 51–70.

Bannerjee-Guha, Swapna, ed. *Accumulation by Dispossession: Transformative Cities in the New Global Order*. New Delhi: Sage, 2010.

Battacharyya, Subhes C. "Review of the Electricity Act 2003 of India." *Dundee Center for Energy, Petroleum and Mineral Law and Policy Online Journal* 14 (2003). http://www.dundee.ac.uk/cepmlp/journal/html/volume14/article14_4.pdf.

Benedict, Ruth. *Patterns of Culture*. New York: Houghton Mifflin, 1934.

Berlant, Lauren. *Cruel Optimism*. Durham, NC: Duke University Press, 2011.

———. "On the Case." *Critical Inquiry* 33 (2007): 663–72.

Béteille, André. *Society and Politics in India: Essays in a Comparative Perspective*. London: Athlone, 1991.

Bhagavan, Manu. "Demystifying the 'Ideal Progressive': Resistance through Mimicked Modernity in Princely Baroda, 1900–1913." *Modern Asian Studies* 35 (2001) 385–409.

———. *The Peacemakers: India and the Quest for One World*. New Delhi: HarperCollins, 2012.

Biehl, João. *Vita: Life in a Zone of Social Abandonment*. Berkeley: University of California Press, 2004.

Blackmur, R. P. *Henry Adams*. Edited by Veronica A. Makowsky. New York: Harcourt, Brace, Jovanovich, 1980.

Blackwood's Edinburgh Magazine. "The Delhi Durbar: A Retrospect." March 1903, 311–23.

Boon, James A. *Affinities and Extremes: Crisscrossing the Bittersweet Ethnology of East Indies History, Hindu-Balinese Culture, and Indo-European Allure*. Chicago: University of Chicago Press, 1990.

——. *Other Tribes, Other Scribes: Symbolic Anthropology in the Comparative Studies of Cultures, Histories, Religions, and Texts*. New York: Cambridge University Press, 1982.

Bornstein, Erica, and Aradhana Sharma. "The Righteous and the Rightful: The Technomoral Politics of NGOs, Social Movements, and the State in India." *American Ethnologist* 43 (2016): 76–90.

Bose, Sugata. "Instruments and Idioms of Colonial and National Development: India's Historical Experience in Comparative Perspective." In *International Development and the Social Sciences: Essays on the History and Politics of Knowledge*, edited by F. Cooper and R. Packard, 45–63. Berkeley: University of California Press, 1997.

Boyer, Dominic. "Anthropology Electric." *Cultural Anthropology* 30 (2015): 531–39.

Broomfield, J. H. *Elite Conflict in a Plural Society: Twentieth-Century Bengal*. Berkeley: University of California Press, 1968.

Cavell, Stanley. *The Claim of Reason: Wittgenstein, Skepticism, Morality, and Tragedy*. Oxford: Clarendon, 1979.

Certeau, Michel de. "History: Science and Fiction." In *Heterologies: Discourse on the Other*, translated by Brian Massumi, 199–221. Minneapolis: University of Minnesota Press, 1986.

Chakrabarty, Dipesh. *Provincializing Europe: Postcolonial Thought and Historical Difference*. 2nd ed. Princeton, NJ: Princeton University Press, 2008.

——. "Reconstructing Liberalism? Notes toward a Conversation between Area Studies and Diasporic Studies." *Public Culture* 10 (1998): 457–81.

Chatterjee, Partha. *Nationalist Thought and the Colonial World: A Derivative Discourse?* London: Zed Books, 1986.

——. *The Politics of the Governed: Reflections on Popular Politics in Most of the World*. New York: Columbia University Press, 2004.

Chaudhuri, Nirad C. *Thy Hand, Great Anarch!* New Delhi: Times Books, 1987.

Choy, Timothy. *Ecologies of Comparison: An Ethnography of Endangerment in Hong Kong*. Durham, NC: Duke University Press, 2011.

Cohn, Bernard S. *Colonialism and Its Forms of Knowledge: The British in India*. Princeton, NJ: Princeton University Press, 1996.

——. "Representing Authority in Victorian India." In *The Invention of Tradition*, edited by E. J. Hobsbawm and T. O. Ranger, 165–209. New York: Cambridge University Press, 1983.

Coleman, Leo. "Ignorance and Government in British India: The Native Fetish of the Crown." In *Regimes of Ignorance*, edited by Roy Dilley and Thomas G. Kirsch, 159–87. New York: Berghahn, 2015.

——. "The Imagining Life: Reflections on Imagination in Political Anthropology." In *Reflections on Imagination: Human Capacity and Ethnographic Method*, edited by Mark Harris and Nigel Rapport, 195–214. Burlington, VT: Ashgate, 2015.

——. "Infrastructure and Interpretation: Meters, Dams, and State Imagination in Scotland and India." *American Ethnologist* 41 (2014): 457–72.

——. "Inside and Outside the House: A Narrative of Mobility and Becoming in Delhi." *Journal of Contemporary Ethnography* (2016). doi:10.1177/0891241616630377.

Collier, Stephen. *Post-Soviet Social: Neoliberalism, Social Modernity, Biopolitics*. Princeton, NJ: Princeton University Press, 2011.

Comaroff, John L., and Jean Comaroff. "Reflections on the Anthropology of Law, Governance, and Sovereignty." In *Rules of Law and Laws of Ruling: On the Governance of Law*, edited by Franz von Benda-Beckmann, Keebet von Benda-Beckmann, and Julia Eckert, 31–60. Burlington, VT: Ashgate, 2009.

Corbridge, Stuart, and John Harriss. *Reinventing India: Liberalization, Hindu Nationalism, and Popular Democracy*. Malden, MA: Blackwell, 2000.

Coronil, Ferdinand. *The Magical State: Nature, Money, and Modernity in Venezuela*. Chicago: University of Chicago Press, 1997.

Creed, Gerald, ed. *The Seductions of Community: Emancipations, Oppressions, Quandaries*. Santa Fe, NM: School of Advanced Research Press, 2006.

Curzon, George Nathaniel. *British Government in India*. 2 vols. London: Cassell, 1925.

——. *Lord Curzon in India; Being a Selection from His Speeches as Viceroy and Governor-General of India, 1898–1905*. Edited by T. Raleigh. London: Macmillan, 1906.

Cyclopedia of India, The. "John Fleming & Co." 289–90. Calcutta: Cyclopedia, 1907.

Dasgupta, Simanti. *Bits of Belonging: Information Technology, Water, and Neoliberal Governance in India*. Philadelphia: Temple University Press, 2015.

Dilks, David. *Curzon in India*. Vol. 1. London: Rupert Hart-Davis, 1969.

Dirks, Nicholas. *Castes of Mind: Colonialism and the Making of Modern India*. Berkeley: University of California Press, 1999.

——. *The Hollow Crown: Ethnohistory of an Indian Kingdom*. New York: Cambridge University Press, 1987.

Douglas, Mary. *Cultural Bias*. London: Royal Anthropological Institute, 1978.

Dubash, Navroz, and Sudhir Chella Rajan. "Power Politics: Process of Power Sector Reform in India." *Economic and Political Weekly*, September 1, 2001.

Dube, Saurabh. *Enchantments of Modernity: Empire, Nation, Globalization*. Chicago: University of Chicago Press, 2009.

——. "Lost and Found: Villages between Anthropology and History." In *Village Matters: Relocating Villages in the Contemporary Anthropology of India*, edited by D. Mines and N. Yazgi, 31–52. New Delhi: Oxford University Press, 2010.

——. "Presence of Europe: An Interview with Dipesh Chakrabarty." *South Atlantic Quarterly* 101 (2002): 859–68.

Dumont, Louis. *Homo Hierarchicus: The Caste System and Its Implications*. Translated by M. Sainsbury, L. Dumont, and B. Gulati. Chicago: University of Chicago Press, 1980.

Dumont, Louis, and David Pocock, eds. *Contributions to Indian Sociology*. Vol. 1. The Hague: Mouton, 1957.

Dupont, Véronique, and U. Ramanathan. "The Courts and the Squatter Settlements in Delhi—or, the Intervention of the Judiciary in Urban 'Governance.'" In *New Forms of Urban Government in India: Shifts, Models, Networks, and Contestations*, edited by I. S. A. Baud and J. de Wit, 312–43. New Delhi: Sage, 2008.

Dutta, Arindam. *The Bureaucracy of Beauty: Design in the Age of Its Global Reproducibility*. New York: Routledge, 2007.

Esposito, Roberto. *Communitas: The Origin and Destiny of Community*. Translated by Timothy Campbell. Stanford, CA: Stanford University Press, 2010.

Eustis, F. A., and Z. H. Zaidi. "King, Viceroy, and Cabinet: The Modification of the Partition of Bengal, 1911." *History* 49 (1964): 171–84.

Evens, T. M. S., and Don Handelman, eds. *The Manchester School: Practice and Ethnographic Praxis in Anthropology*. New York: Berghahn, 2006.

Fernandes, Leela. *India's New Middle Class: Democratic Politics in an Era of Economic Reform*. Minneapolis: University of Minnesota Press, 2006.

Fischer, Michael M. J. "Four Genealogies for a Recombinant Anthropology of Science and Technology." *Cultural Anthropology* 22 (2007): 539–615.

Frasch, Tilman. "Tracks in the City: Technology, Mobility, and Society in Colonial Rangoon and Singapore." *Modern Asian Studies* 46 (S01) (1912): 97–118.

Frykenberg, R. E. "The Coronation Durbar of 1911: Some Implications." In *Delhi through the Ages: Selected Essays in Urban History, Culture, and Society*, 225–46. Delhi: Oxford University Press, 1986.

Fuller, C. J., and Véronique Bénéï, eds. *Everyday State and Society in India*. London: Hurst, 2001.

Fustel de Coulanges, Numa Denis. *The Ancient City: A Study on the Religion, Laws, and Institutions of Greece and Rome*. Baltimore: Johns Hopkins University Press, 1980. Original: *La cité antique: Étude sur le culte, le droit, les institutions de la Grèce et de Rome*. 11. éd. Paris: Hachette, 1885.

Gandhi, Leela. *Affective Communities: Anticolonial Thought, Fin-de-Siècle Radicalism, and the Politics of Friendship*. Durham, NC: Duke University Press, 2006.

Gandhi, M. K. *"Hind Swaraj" and Other Writings*. Edited by Anthony J. Parel. New York: Cambridge University Press, 1997.

Garces, Chris. "The Cross Politics of Ecuador's Penal State." *Cultural Anthropology* 25 (2010): 459–96.

Geertz, Clifford. "Centers, Kings, and Charisma: Reflections on the Symbolics of Power." In *Local Knowledge: Further Essays in Interpretive Anthropology*, 121–46. New York: Basic Books, 1983.

——. *Negara: The Theatre State in Nineteenth-Century Bali*. Princeton, NJ: Princeton University Press, 1980.

——. *The Social History of an Indonesian Town*. Cambridge, MA: MIT Press, 1965.

George, King of Great Britain. *Speeches of His Majesty King George in India*. 2nd ed. Madras: G. A. Natesan, 1912.

Ghertner, D. Asher. "Calculating without Numbers: Aesthetic Governmentality in Delhi's Slums." *Economy and Society* 39 (2010): 185–217.

———. "Nuisance Talk and the Propriety of Property: Middle-Class Discourses of a Slum-Free Delhi." *Antipode* 44 (2012): 1161–87.

Gilman, Nils. *Mandarins of the Future: Modernization Theory in Cold-War America*. Baltimore: Johns Hopkins University Press, 2007.

Gilmour, David. *Curzon*. London: John Murray, 1994.

Glover, William. *Making Lahore Modern: Constructing and Imagining a Colonial City*. Minneapolis: University of Minnesota Press, 2008.

Goswami, Manu. *Producing India: From Colonial Economy to National Space*. Chicago: University of Chicago Press, 2004.

Greenhouse, Carol J. Introduction to *Ethnographies of Neoliberalism*, edited by Carol J. Greenhouse, 1–11. Philadelphia: University of Pennsylvania Press, 2010.

———. *The Paradox of Relevance: Ethnography and Citizenship in the United States*. Philadelphia: University of Pennsylvania Press, 2011.

Gupta, Akhil. "Blurred Boundaries: The Discourse of Corruption, the Culture of Politics, and the Imagined State." *American Ethnologist* 22 (1995): 375–402.

———. *Red Tape: Bureaucracy, Structural Violence, and Poverty in India*. Durham, NC: Duke University Press, 2012.

Gupta, Narayani. *Delhi between Two Empires, 1803–1931: Society, Government, and Urban Growth*. Delhi: Oxford University Press, 1981.

Gupta, S. K. D. "Review of Legislation 1947: Eastern Countries, India—Central Legislature." *Journal of Comparative Legislation and International Law*, 3rd series, 31 (1949): 116–19.

Handelman, Don. *Models and Mirrors: Towards an Anthropology of Public Events*. 2nd ed. Oxford: Berghahn, 1990.

Hannah, Leslie. *Electricity before Nationalisation: A Study of the Development of the Electricity Supply Industry in Britain to 1948*. Baltimore: Johns Hopkins University Press, 1979.

Hansen, Thomas Blom. *The Saffron Wave: Democracy and Hindu Nationalism in Modern India*. Princeton, NJ: Princeton University Press, 1999.

Hansen, Thomas Blom, and Oskar Verkaaik. "Urban Charisma: On Everyday Mythologies in the City." *Critique of Anthropology* 29 (2009): 5–26.

Harriss, John. "'Politics Is a Dirty River': But Is There a 'New Politics' of Civil Society? Perspectives from Global Cities of India and Latin America." In *Power Matters: Essays on Institutions, Politics, and Society in India*, 256–74. New Delhi: Oxford University Press, 2006.

Harvey, Penny, and Hannah Knox. *Roads: An Anthropology of Infrastructure and Expertise*. Ithaca, NY: Cornell University Press, 2015.

Haynes, Douglas E. *Rhetoric and Ritual in Colonial India: The Shaping of a Public Culture in Surat City, 1852–1928*. Berkeley: University of California Press, 1991.

Heesterman, J. C. *The Inner Conflict of Tradition: Essays in Indian Ritual, Kingship, and Society*. Chicago: University of Chicago Press, 1985.

Hess, David. "Electricity Transformed: Neoliberalism and Local Energy in the United States." *Antipode* 43 (2011): 1056–77.

Hobsbawm, E. J., and T. O. Ranger, eds. *The Invention of Tradition*. New York: Cambridge University Press, 1983.

Hocart, A. M. *Kings and Councillors: An Essay in the Comparative Anatomy of Human Society*. Chicago: University of Chicago Press, 1970.

———. *Kingship*. 1927. Reprinted ed., Oxford: Oxford University Press, 1969.

Hughes, Thomas Parke. *Networks of Power: Electrification in Western Society, 1880–1930*. Baltimore: Johns Hopkins University Press, 1983.

Hull, Matthew S. "Communities of Place, Not Kind: American Technologies of Neighborhood in Postcolonial Delhi." *Comparative Studies in Society and History* 53 (2011): 757–90.

Hussain, Nazia. "The City of Dacca, 1921–1947: Society, Water, and Electricity." In *The City in South Asia: Pre-Modern and Modern*, edited by K. Ballhatchet and J. Harrison, 197–223. London: Curzon, 1980.

Inden, Ronald. "Embodying God: From Imperial Progresses to National Progress in India." *Economy and Society* 24 (1995): 245–78.

Indian National Congress. *Power and Fuel: Report of the Subcommittee of the National Planning Committee*. Edited by K. T. Shah. Bombay: Vora, 1949.

Irving, Robert Grant. *Indian Summer: Lutyens, Baker, and Imperial Delhi*. New Haven, CT: Yale University Press, 1981.

Jameson, Fredric. *A Singular Modernity: Essay on the Ontology of the Present*. New York: Verso, 2002.

Jasanoff, Sheila. *Designs on Nature: Science and Democracy in Europe and the United States*. Princeton, NJ: Princeton University Press, 2005.

Johnson, David. "A British Empire for the Twentieth Century: The Inauguration of New Delhi, 1931." *Urban History* 35 (2008): 462–84.

———. "Land Acquisition, Landlessness, and the Building of New Delhi." *Radical History Review* 108 (2010): 91–116.

Joyce, Patrick. "Filing the Raj: Political Technologies of the Imperial British State." In *Material Powers: Cultural Studies, History, and the Material Turn*, edited by Tony Bennett and Patrick Joyce, 102–23. New York: Routledge, 2010.

Kale, Sunila S. "Current Reforms: The Politics of Policy Change in India's Electricity Sector." *Pacific Affairs* 77 (2004): 467–91.

———. *Electrifying India: Regional Political Economies of Development*. Stanford, CA: Stanford University Press, 2014.

Kalyan, Rohan. "The Magician's Ghetto: Moving Slums and Everyday Life in a Postcolonial City." *Theory, Culture & Society* 31 (2014): 49–73.

Kamath, Lalitha, and M. Vijayabaskar. "Limits and Possibilities of Middle Class Associations as Urban Collective Actors." *Economic and Political Weekly* 44 (2009): 368–76.

Kapferer, Bruce. "Situations, Crisis, and the Anthropology of the Concrete: The Contribution of Max Gluckman." In *The Manchester School: Practice and Ethnographic Praxis in Anthropology*, edited by T. M. S. Evens and Don Handelman, 118–58. New York: Berghahn, 2006.

Kaviraj, Sudipta. "On the Enchantment of the State: Indian Thought on the Role of the State in the Narrative of Modernity." *European Journal of Sociology* 46 (2005): 263–96.

Keith, Arthur Berriedale. *The King and the Imperial Crown: The Powers and the Duties of His Majesty.* London: Longmans, Green, 1936.

Kosambi, Meera. *Bombay in Transition: The Growth and Social Ecology of a Colonial City.* Stockholm: Almqvist & Wiksell, 1986.

Kracauer, Siegfried. "The Mass Ornament." In *The Mass Ornament: Weimar Essays.* Translated by Thomas Y. Levin, 75–86. Cambridge: Harvard University Press.

Larkin, Brian. "The Politics and Poetics of Infrastructure." *Annual Review of Anthropology* 42 (2013): 327–43.

Laski, Harold. *Parliamentary Government in England.* New York: Viking, 1938.

Latour, Bruno. "Morality and Technology: The End of the Means." Translated by Couze Venn. *Theory, Culture & Society* 19 (2002): 247–60.

———. *Reassembling the Social: An Introduction to Actor-Network Theory.* New York: Oxford University Press, 2005.

Legg, Stephen. *Spaces of Colonialism: Delhi's Urban Governmentality.* Malden, MA: Blackwell, 2007.

Mann, Michael. "The Autonomous Power of the State." *Archives Européennes de Sociologie* 25 (1984): 185–213.

Mann, Michael. "Torchbearers on the Path of Progress: Britain's Ideology of 'Moral and Material Progress' in India; An Introductory Essay." In *Colonialism as Civilizing Mission: Cultural Ideology in British India,* edited by Harald Fischer-Tiné and Michael Mann, 1–29. London: Anthem, 2004.

Mantena, Karuna. *Alibis of Empire: Henry Maine and the Ends of Liberal Imperialism.* Princeton, NJ: Princeton University Press, 2010.

Marriott, McKim. "Little Communities in an Indigenous Civilization." In *Village India: Studies in the Little Community,* edited by McKim Marriott, 171–222. Chicago: University of Chicago Press, 1955.

———. 1963. Review of *Gopalpur: An Indian Village,* by Alan Beals. *American Anthropologist* 65, no. 6: 1366–69.

———, ed. *Village India: Studies in the Little Community.* Chicago: University of Chicago Press, 1955.

Mazzarella, William. "Internet X-Ray: E-Governance, Transparency, and the Politics of Immediation in India." *Public Culture* 18 (2006): 473–505.

———. *Shoveling Smoke: Advertising and Globalization in Contemporary India.* Durham, NC: Duke University Press, 2003.

Mehta, Uday Singh. "The Social Question and the Problem of History after Empire." In *Lineages of Empire: The Historical Roots of British Imperial Thought,* edited by Duncan Kelly, 31–49. New York: Oxford University Press, 2009.

Mehta, Ved. *Portrait of India.* New York: Farrar, Straus & Giroux, 1970.

Menpes, Mortimer, and Dorothy Menpes. *The Durbar.* London: A. & C. Black, 1903.

Metcalf, Thomas. *Ideologies of the Raj.* New York: Cambridge University Press, 1994.

Middleton, Townsend. "Anxious Belongings: Anxiety and the Politics of Belonging in Subnationalist Darjeeling." *American Anthropologist* 115 (2013): 608–21.

Milbert, Isabelle. "Law, Urban Policies, and the Role of Intermediaries in Delhi." In *New Forms of Urban Governance in India: Shifts, Models, Networks, and Contestations*, edited by I. S. A. Baud and J. de Wit, 177–212. New Delhi: Sage, 2008.

Miller, Daniel. Introduction in *Materiality*, edited by Daniel Miller, 1–49. Durham, NC: Duke University Press, 2005.

Mines, Diane P., and Nicolas Yazgi, eds. *Village Matters: Relocating Villages in the Contemporary Anthropology of India*. New Delhi: Oxford University Press, 2010.

Mitchell, Timothy. *Carbon Democracy: Political Power in the Age of Oil*. New York: Verso, 2011.

Mitra, Asok. *Delhi: Capital City*. Delhi: Thomson Press (India), 1970.

Moore, Sally Falk, and Barbara G. Myerhoff, eds. *Secular Ritual*. Assen, Netherlands: Van Gorcum, 1977.

Mrázek, Rudolf. *Engineers of Happy Land: Technology and Nationalism in a Colony*. Princeton, NJ: Princeton University Press, 2002.

Naipaul, V. S. *India: A Wounded Civilization*. New York: Knopf, 1977.

Nehru, Jawaharlal. *The Discovery of India*. Centenary edition. New Delhi: Oxford University Press, 1985.

——. *Glimpses of World History*. New Delhi: Oxford University Press, 1989.

Nilsson, Sten. *The New Capitals of India, Pakistan, and Bangladesh*. Translated by Elisabeth Andréasson. Lund, Sweden: Studentlitteratur, 1973.

Oldenburg, Philip. *Big City Government in India: Councilor, Administrator, and Citizen in Delhi*. Tucson: University of Arizona Press, 1976.

Parry, Jonathan. "Nehru's Dream and the Village 'Waiting Room': Long-Distance Migrants to an Indian Steel Town." *Contributions to Indian Sociology*, n.s., 37 (2003): 219–49.

——. "The Sacrifices of Modernity in a Soviet-Built Steel Town in Modern India." In *On the Margins of Religion*, edited by Frances Pine and João de Pina-Cabral, 233–62. Oxford: Berghahn, 2008.

Peel, George. "At the Durbar." *Cornhill Magazine* 14, n.s. (1903): 309–18.

Pioneer Press. *The Coronation Durbar at Delhi*. Allahabad: Pioneer Press, 1903.

Pothen, Nayantara. *Glittering Decades: New Delhi in Love and War*. New Delhi: Penguin, 2012.

Prakash, Gyan. *Another Reason: Science and the Imagination of Modern India*. Princeton, NJ: Princeton University Press, 1999.

——. Introduction to *The Spaces of the Modern City*, edited by Gyan Prakash and Kevin M. Kruse, 1–18. Princeton, NJ: Princeton University Press, 2008.

Prasad, Ritika. *Tracks of Change: Railways and Everyday Life in India*. Delhi: Cambridge University Press, 2015.

Qureshi, Ishtiaq Husain. *The Administration of the Sultanate of Delhi*. Lahore: Sh. M. Ashraf, 1944.

Rabinow, Paul. *Marking Time: On the Anthropology of the Contemporary*. Princeton, NJ: Princeton University Press, 2007.

Raghavan, G. N. S., ed. *M. Asaf Ali's Memoirs: The Emergence of Modern India*. Delhi: Ajanta, 1994.

Raleigh, T., ed. *Lord Curzon in India; Being a Selection from His Speeches as Viceroy and Governor-General of India, 1898–1905*. London: Macmillan, 1906.

Rancière, Jacques. *The Politics of Aesthetics: The Distribution of the Sensible*. Translated by Gabriel Rockhill. New York: Continuum, 2004.

Rao, Srinivasa, and John Lourdusamy. "Colonialism and the Development of Electricity: The Case of Madras Presidency, 1900–1947." *Science, Technology & Society* 15 (2010): 27–54.

Redfield, Robert, and Milton Singer. "The Cultural Role of Cities." *Economic Development and Cultural Change* 3 (1954): 53–73.

Reed, Stanley. *The King and Queen in India*. Bombay: Bennett, Coleman, 1912.

Riles, Annelise. *Collateral Knowledge: Legal Reasoning in Global Financial Markets*. Chicago: University of Chicago Press, 2011.

Robbins, Bruce. "The Smell of Infrastructure: Notes toward an Archive." *boundary 2*, no. 34 (2007): 26–33.

Rogers, Douglas. *The Depths of Russia: Oil, Power, and Culture after Socialism*. Ithaca, NY: Cornell University Press, 2015.

Roy, Parama. *Alimentary Tracts: Appetites, Aversions, and the Postcolonial*. Durham, NC: Duke University Press, 2010.

Roy, Srirupa. *Beyond Belief: India and the Politics of Post-colonial Nationalism*. Durham, NC: Duke University Press, 2007.

Rubenstein, Michael D. *Public Works: Infrastructure, Irish Modernism, and the Postcolonial*. South Bend, IN: Notre Dame University Press, 2010.

Ruet, Joël. *Privatising Power Cuts? Ownership and Reform of the State Electricity Boards in India*. New Delhi: Academic Foundation / Centre des Sciences Humaines, 2005.

Ruet, Joël, and Stéphanie Tawa Lama-Rewal, eds. *Governing India's Metropolises*. New Delhi: Routledge, 2009.

Sagar, Jagdish. "Roundtable on Power Sector Reforms." *IIMB Management Review*, March 2004, 71–84.

Saha, Meghnad. *Collected Works of Meghnad Saha*. Vol. 2. Edited by Santimay Chatterjee. Calcutta: Saha Institute / Orient Longman, 1986.

Sahlins, Marshall. "Infrastructuralism." *Critical Inquiry* 36 (2010): 371–85.

Said, Edward. *Orientalism*. New York: Pantheon, 1978.

Santhakumar, V. *Analysing Social Opposition to Reforms: The Electricity Sector in India*. New Delhi: Sage, 2008.

Saran, Rohit. "How We Live: Census of India Household Survey." *India Today* (international edition), July 28, 2003.

Sartori, Andrew. "Robert Redfield's Comparative Civilizations Project and the Political Imagination of Postwar America." *positions* 6 (1998): 33–65.

Sassen, Saskia. *Territory, Authority, Rights: From Medieval to Global Assemblages*. Princeton, NJ: Princeton University Press, 2006.

Scheppele, Kim Lane. "Constitutional Ethnography: An Introduction." *Law and Society Review* 38 (2004): 389–406.

Schivelbusch, Wolfgang. *Disenchanted Night: The Industrialisation of Light in the Nineteenth Century*. Translated by Angela Davies. New York: Berg, 1988.

Scott, J. Barton, and Brannon D. Ingram. "What Is a Public? Notes from South Asia." *South Asia: Journal of South Asian Studies* 38 (2015): 357–70.

Sen, Dhirendranath. *From Raj to Swaraj*. Calcutta: Vidyodaya, 1954.

Sethi, Aman. "A Shock to Reformers / The Price of Reforms." *Frontline* (India) 22 (19) (September 10–23, 2005). frontline.in.

Shamir, Ronen. *Current Flow: The Electrification of Palestine*. Stanford, CA: Stanford University Press, 2013.

Sharma, Aradhana. "Epic Fasts and Shallow Spectacles: The 'India against Corruption' Movement, Its Critics, and the Re-making of Gandhi." *South Asia: The Journal of South Asian Studies* 37 (2014): 365–80.

Shils, Edward. "Charisma, Order, and Status." *American Sociological Review* 30 (1965): 199–213.

Singer, Milton. *When a Great Tradition Modernizes: An Anthropological Approach to Indian Civilization*. New York: Praeger, 1972.

Singh, Sangat. *Freedom Movement in Delhi, 1858–1919*. New Delhi: Associated Publishing House, 1972.

Sivaramakrishnan, K. C., ed. *People's Participation in Urban Governance*. New Delhi: Institute of Social Sciences, 2006.

Skaria, Ajay. "Relinquishing Republican Democracy: Gandhi's Ramarajya." *Postcolonial Studies* 14 (2011): 203–29.

Sklair, Leslie. *The Sociology of Progress*. Boston: Routledge & Kegan Paul, 1970.

Spickard, James V. "A Guide to Mary Douglas's Three Version of Grid/Group Theory." *Sociological Analysis* 50 (1989): 151–70.

Spivak, Gayatri. *A Critique of Postcolonial Reason: Toward a History of the Vanishing Present*. Cambridge, MA: Harvard University Press, 1999.

Srinivas, M. N. "The Dominant Caste in Rampura." *American Anthropologist*, n.s., 61 (1959): 1–16.

——. "A Note on Mr. Goheen's Note: Forum on India's Cultural Values and Economic Development." *Economic Development and Cultural Change* 7 (1958): 3–6.

——. *The Remembered Village*. Berkeley: University of California Press, 1976.

Srivastava, Sanjay. "Urban Spaces, Disney-Divinity, and Moral Middle Classes in Delhi." *Economic and Political Weekly* 44, nos. 26 & 27 (2009): 338–45.

Steer, Valentia. *The Delhi Durbar, 1902–1903*. Madras: Higginbotham, 1903.

Steinmetz, George. *The Devil's Handwriting: Precoloniality and the German Colonial State in Qingdao, Samoa, and Southwest Africa*. Chicago: University of Chicago Press, 2007.

Stengers, Isabelle. *Thinking with Whitehead: A Free and Wild Creation of Concepts*. Translated by Michael Chase. Cambridge, MA: Harvard University Press, 2011.

Stokes, Eric. *The English Utilitarians and India*. Oxford: Clarendon, 1959.

Stoler, Ann. *Along the Archival Grain: Epistemic Anxieties and Colonial Common Sense*. Princeton, NJ: Princeton University Press, 2009.

Sundaram, Ravi. *Pirate Modernity: Delhi's Media Urbanism*. New York: Routledge, 2009.

Sunder Rajan, Rajeswari. *The Scandal of the State: Women, Law, and Citizenship in Postcolonial India*. Durham, NC: Duke University Press, 2003.

Tarlo, Emma. *Unsettling Memories: Narratives of the Emergency in Delhi*. Berkeley: University of California Press, 2003.

Taylor, Charles. *A Secular Age*. Cambridge, MA: Belknap Press of Harvard University Press, 2007.

Tinker, Hugh. *The Foundations of Local Self-Government in India, Pakistan, and Burma*. London: Athlone, 1954.

Trevithick, Alan. "Some Structural and Sequential Aspects of the British Imperial Assemblages at Delhi: 1877–1911." *Modern Asian Studies* 24 (1990): 561–78.

Truelove, Yaffa, and Emma Mawdsley. "Discourses of Citizenship and Criminality in Clean, Green Delhi." In *A Companion to the Anthropology of India*, edited by Isabelle Clark-Decès, 407–25. Malden, MA: Blackwell, 2011.

Vajpeyi, Ananya. *Righteous Republic: The Political Foundations of Modern India*. Cambridge, MA: Harvard University Press, 2012.

Valverde, Mariana. "Seeing Like a City: The Dialectic of Modern and Premodern Ways of Seeing in Urban Governance." *Law & Society Review* 45 (2011): 277–312.

Visvanathan, Shiv. "Between Cosmology and System: The Heuristics of a Dissenting Imagination." In *Another World Is Possible: Beyond Northern Epistemologies*, edited by Bonaventura de Sousa Santos, 182–218. New York: Verso, 2008.

Von Schnitzler, Antina. "Traveling Technologies: Infrastructure, Ethical Regimes, and the Materiality of Politics in South Africa." *Cultural Anthropology* 28 (2013): 670–93.

Wagner, Roy. *The Invention of Culture*. Rev. and expanded ed. Chicago: University of Chicago Press, 1981.

Waldrop, Anne. "Gating and Class Relations: The Case of a New Delhi 'Colony.'" *City and Society* 16, no. 2 (2004): 93–116.

Webb, Martin. "Short Circuits: The Aesthetics of Protest, Media, and Martyrdom in Indian Anti-corruption Activism." In *Political Aesthetics*, edited by P. Werbner, M. Webb, and K. Spellman-Poots, 193–221. Edinburgh: University of Edinburgh Press, 2014.

Weber, Max. "Class, Status, Party." In *From Max Weber: Essays in Sociology*, translated by H.H. Geerth and C. Wright Mills, 180–95. New York: Oxford University Press, 1946.

——. *Economy and Society*. Vol. 1. Edited by Guenther Roth and Claus Wittich. New York: Bedminister, 1968.

Wheeler, Stephen. *History of the Delhi Coronation Durbar*. London: J. Murray, 1904.

Whitehead, Alfred North. *Symbolism: Its Meaning and Effect*. New York: Capricorn Books, 1959. First published in 1927 by Macmillan.

——. *Process and Reality: An Essay in Cosmology*, Corr. Ed. Edited by David Ray Griffin and Donald W. Sherburne. New York: Free Press, 1978. First published in 1929 by Macmillan.

Williams, Rosalind H. *Notes on the Underground: An Essay on Technology, Society, and the Imagination*. New ed. Cambridge, MA: MIT Press, 2008.

———. "Our Technological Age, from Inside Out." *Technology and Culture* 55 (2014): 461–76.

Zachariah, Benjamin. *Developing India: An Intellectual and Social History, c. 1930–1950.* New Delhi: Oxford University Press, 2005.

Zérah, Marie-Hélène. "Reconfiguring Power Relations: Policies toward Urban Services in Mumbai." In *Accumulation by Dispossession: Transformative Cities in the New Global Order*, edited by Swapna Bannerjee-Guha, 151–68. New Delhi: Sage, 2010.

Zetland, Lawrence John Lumley Dundas, Marquis of. *The Life of Lord Curzon: Being the Authorized Biography of George Nathaniel, Marquess Curzon of Kedleston, K.G., by the Rt. Hon. the Earl of Ronaldshay*. London: E. Benn, 1928.

INDEX

Note: Italic page numbers refer to illustrations.

CPSIA information can be obtained
at www.ICGtesting.com
Printed in the USA
FFOW03n2017140417
34483FF